HIGH-TECHNOLOGY CLUSTERS, NETWORKING AND COLLECTIVE LEARNING IN EUROPE

High-Technology Clusters, Networking and Collective Learning in Europe

Edited by

DAVID KEEBLE and FRANK WILKINSON
ESRC Centre for Business Research, University of Cambridge

On behalf of the TSER European Network

LONDON AND NEW YORK

First published 2000 by Ashgate Publishing

Reissued 2018 by Routledge
2 Park Square, Milton Park, Abingdon, Oxon OX14 4RN
711 Third Avenue, New York, NY 10017, USA

Routledge is an imprint of the Taylor & Francis Group, an informa business

Copyright © David Keeble and Frank Wilkinson 2000

All rights reserved. No part of this book may be reprinted or reproduced or utilised in any form or by any electronic, mechanical, or other means, now known or hereafter invented, including photocopying and recording, or in any information storage or retrieval system, without permission in writing from the publishers.

Notice:
Product or corporate names may be trademarks or registered trademarks, and are used only for identification and explanation without intent to infringe.

Publisher's Note
The publisher has gone to great lengths to ensure the quality of this reprint but points out that some imperfections in the original copies may be apparent.

Disclaimer
The publisher has made every effort to trace copyright holders and welcomes correspondence from those they have been unable to contact.

A Library of Congress record exists under LC control number: 00132849

ISBN 13: 978-1-138-73160-8 (hbk)
ISBN 13: 978-1-138-73158-5 (pbk)
ISBN 13: 978-1-315-18890-4 (ebk)

Contents

List of Figures vii
List of Tables and Appendices ix
List of Contributors xi

1 **High-Technology** SMEs, Regional Clustering and Collective Learning: An Overview 1
   David Keeble and Frank Wilkinson

2 High-Technology Clusters and Evolutionary Trends in the 1990s 21
   Christian Longhi and David Keeble

3 Regional Institutional and Policy Frameworks for High-Technology SMEs in Europe 57
   Christine Tamásy and Rolf Sternberg

4 University and Public Research Institute Links with Regional High-Technology SMEs 90
   Helen Lawton Smith and Michel de Bernardy

5 The Role of Inter-SME Networking and Links in Innovative High-Technology Milieux 118
   Roberto Camagni and Roberta Capello

6 Large Firm Acquisitions, Spin-Offs and Links in the Development of Regional Clusters of Technology-Intensive SMEs 156
   Åsa Lindholm Dahlstrand

7 Collective Learning, System Competences and Epistemically Significant Moments 182
   Clive Lawson

8	Collective Learning Processes in European High-Technology Milieux *David Keeble*	199
9	Concluding Reflections: Some Policy Implications *Frank Wilkinson and Barry Moore*	230

Index 260

List of Figures

Figure 2.1	Growth, decline and resurgence in the UK's small firm population, 1980–97	26
Figure 2.2	Europe's leading science and technology regions, 1995	32
Figure 2.3	Exogenous and endogenous employment growth in Sophia-Antipolis, 1991–97	40
Figure 2.4	Regional clustering and the European entrepreneurial life sciences sector, 1999	45
Figure 3.1	Impacts of regional institutional frameworks on high-technology SMEs' performance	59
Figure 3.2	Impacts of technology policies on high-technology SMEs' performance	61
Figure 3.3	EU R&D expenditure by region	63
Figure 3.4	European regional clusters of high-technology SMEs: towards a typology	70
Figure 4.1	Relations between firms and knowledge institutions: a new paradigm is supplanting the old pipeline one	113
Figure 5.1a	Italian high-technology SMEs: local and external suppliers	124
Figure 5.1b	Italian high-technology SMEs: local and external customers	124
Figure 5.2	Innovative linkages of innovative manufacturing SMEs in the Hannover-Brunswick-Gottingen Research Triangle, by firm size and co-operation partner	129
Figure 5.3a	Italian high-technology SMEs: the importance of local suppliers for firms' innovative activity	130
Figure 5.3b	Italian high-technology SMEs: the importance of local customers for firms' innovative activity	130
Figure 5.4	The importance of local inter-firm links to Cambridge technology-intensive SMEs	133
Figure 5.5	High-technology SMEs in the Utrecht region: the importance of innovation partners	134

Figure 5.6a   High-technology SMEs in the Utrecht region: the
              importance of regional customers   135
Figure 5.6b   High-technology SMEs in the Utrecht region: the
              importance of regional suppliers   135
Figure 5.7   Factor productivity with respect to local inter-SME links 139
Figure 6.1   An economic system of ownership changes   160

List of Tables and Appendices

Table 1.1	High-technology industries in the United Kingdom	4
Table 2.1	New or enhanced characteristics of organisation of production by high-technology SMEs in Great Britain	47
Table 3.1	Assessment of institutional frameworks for high-technology SMEs in 10 European regions	65
Table 3.2	Government budget appropriations or outlays for R&D categorised by socio-economic objectives (1996)	68
Appendix 3.1	Selected characteristics of regional innovation systems	75
Table 4.1	Typology of regional influences	93
Table 5.1	Functions of the local milieu	123
Table 5.2	European high-technology milieux analysed	125
Table 5.3a	Local inter-firm links in the Cambridge region	126
Table 5.3b	The importance of local inter-firm links in the Cambridge region	126
Table 5.4	The number and importance of different inter-firm links in Oxfordshire	127
Table 5.5	Sources of innovation reported as moderately to highly significant by firms in Oxfordshire	128
Table 5.6	Results of a cluster analysis of learning behaviour in the three Italian milieux	132
Table 5.7	The importance of local links for firms' development in Oxfordshire	132
Table 5.8	Results from regression analyses	140
Appendix 5.1	Factor and cluster analyses	145
Table 5.9a	Factor analysis of the structural and innovative characteristics of firms	147
Table 5.9b	Factor analysis of suppliers' relationships	148
Table 5.9c	Factor analysis of customers' relationships	149
Table 5.9d	Factor analysis of location advantages	150
Table 5.9e	Channels of knowledge acquisition	151
Appendix 5.2	The production function model	152
Table 5.10	Results of the econometric estimation	154

Table 6.1	Large firms and categories of innovative regions	169
Table 6.2	Sectoral and technological specialisation	170
Table 6.3	Entrepreneurial spin-offs	171
Table 6.4	Local and external links between SMEs and parents or other large firms	173
Table 6.5	Primary large firm acquirers	175
Table 8.1	Pre-conditions for learning: examples of regional collective agents	205
Table 8.2	Technology-intensive spin-offs as a regional collective learning process	208
Table 8.3	Research and managerial staff recruitment and mobility within the Cambridge region	211
Table 8.4	Regional labour market recruitment and flows of managerial and technological expertise	211
Table 8.5	Inter-firm links and networks in regional knowledge sharing and collective learning	215
Table 8.6	Knowledge centres and regional collective learning: European examples	222
Table 8.7	New regional collective initiatives in the Cambridge region, 1997–99	224

List of Contributors

Roberto Camagni is Professor of Economics and Urban Economics at the Politecnico di Milano. He has worked extensively on innovation diffusion theory and on urban and regional economics. He is President of GREMI and has been Chairman of the Italian Regional Science Association. During the Prodi Government he was Head of the Department for Urban Affairs at the Presidency of the Council of Ministers. His textbook on Urban Economics, in Italian, has been translated into French and is forthcoming in Spanish.

Roberta Capello is an associate Professor of Regional and Urban Economics at the University of Molise, and of Economics at the Politecnico of Milan. She has been the Secretary of the Italian Section of the Regional Science Association International and is now Member of the Organising Committee of the European Regional Science Association and Treasurer of this association. She has worked extensively on innovation diffusion theory and on urban and regional economics.

Michel de Bernardy is a researcher in Geography associated with the Laboratoire de la Montagne Alpine at the Joseph Fourier University, Grenoble. A longstanding member of GREMI, his main research areas are innovative processes and adaptive and endogenous local capabilities for sustainable development.

David Keeble is Lecturer in Economic Geography, Fellow of St Catharine's College, and former Assistant Director and SME Programme Director of the ESRC Centre for Business Research at Cambridge University. His research interests focus on the geography of new firm formation, regional and urban-rural variations in small business innovation and growth, and the growth of high-technology clusters, in Britain and the European Union.

Clive Lawson is a Fellow of Girton College, Cambridge, and Research Associate at Cambridge University's ESRC Centre for Business Research. He has recently worked on regional learning and networking, and various issues relating to economic methodology.

Helen Lawton Smith is Reader in Local Economic Development, Centre for Local Economic Development, Coventry University. She was previously Research Director (1992–95) then Director of Science Policy Studies (1992–98) at the Regulatory Policy Research Centre, Hertford College, Oxford. She is Senior Research Associate at the School of Geography, Oxford University and a Research Associate at the Centre for Business Research, University of Cambridge. Her research interests include the geography of innovation and the governance of innovation systems.

Åsa Lindholm Dahlstrand is Associate Professor at the Department of Industrial Dynamics, Chalmers University of Technology, Göteborg, Sweden. She works in the field of technology management and her research focuses on small technology-based firms and the role of entrepreneurs in the development of firms and economies. She is especially interested in the study of acquisitions and spin-offs as various entrepreneurial forms of promoting innovativeness and growth.

Christian Longhi is a Senior Researcher at the IDEFI-LATAPSES (CNRS and University of Nice-Sophia-Antipolis). His main research interests cover the fields of innovative strategies of firms and their impact on local development, of regional and national patterns of innovation, and of knowledge infrastructures. His empirical analysis has been applied to an economic understanding of science and technology parks (the 'technopolis' phenomenon).

Barry Moore is Assistant Director of Research in the Department of Land Economy, Cambridge University and Fellow in Economics of Downing College. His research interests focus on urban and regional policy evaluation in the UK and Europe Union, and the growth of innovative high-technology SMEs and university-industry relationships.

Rolf Sternberg is Professor of Economic Geography and Head of the Department of Economic and Social Geography, Faculty of Economics and Social Sciences, University of Cologne. His research focuses on the consequences of technological change for regional development, which he analyses theoretically as well as empirically. He has also published extensively on the implications of national, regional and local technology policies for regional development.

Christine Tamásy is researcher and lecturer at the Department of Economic and Social Geography of the University of Cologne. Her fields of research are industrial geography, entrepreneurship research (especially innovation centres), and sustainable regional development.

Frank Wilkinson is a Senior Research Officer in the University of Cambridge, Department of Applied Economics and the ESRC Centre for Business Research, and a Life Fellow of Girton College, Cambridge. His research is concerned mainly with the effects of industrial and labour institutions and organisations on economic performance.

1 High-Technology SMEs, Regional Clustering and Collective Learning: An Overview

DAVID KEEBLE AND FRANK WILKINSON

1.1 Introduction, Aims and Objectives

This book brings together and synthesises findings from original research by a group of leading European researchers into the recent evolution and dynamic processes underpinning the growth of key European regional clusters of technology-based small and medium-sized enterprises (SMEs). The growth of these regional clusters of technology-intensive firms, as exemplified in such regions as Cambridge, Oxford, Grenoble, Sophia-Antipolis, Munich and Goteborg, represents one of the most fascinating and arguably significant examples of the emergence since the 1970s of what have been termed 'new industrial spaces' (Scott, 1988; Storper, 1993, 1995) and 'innovative milieux' (Aydalot, 1986; Aydalot and Keeble, 1988a; Camagni, 1991a; Ratti et. al., 1997) in both previously non-industrial areas and through the restructuring of existing metropolitan and industrial regions. While large firms are often involved in the growth of these regional high-technology clusters, all are characterised by substantial numbers of small, new and innovative enterprises engaged in technologically-advanced manufacturing and service activities. Notwithstanding the impact of the early-1990s European-wide recession, most of these clusters appear to have been growing rapidly in the 1990s, through processes such as new firm spin-off and endogenous expansion: and many observers have suggested that they are characterised by new forms of production organisation, based on high levels of inter-firm collaboration and cooperation, strong links with local knowledge centres such as universities, and the development of a regionally-embedded capacity for 'collective learning' and technology development linked to the growing local concentration of scientific, technological and managerial expertise.

The growth of key examples of these regional clusters, and the processes involved, have been studied in detail by members of the TSER European research network on 'Networks, Collective Learning and RTD (Research and Technology Development) in Regionally-Clustered High-Technology Small and Medium-Sized Enterprises'. Meetings of this network have been funded by Directorate-General XII for Science, Research and Development of the European Commission, under the Targeted Socio-Economic Research (TSER) Initiative of the Fourth Framework Programme, whose financial support is gratefully acknowledged. The regions studied and the researchers involved are: Cambridge (David Keeble, Clive Lawson, Barry Moore, Frank Wilkinson and Elisabeth Garnsey), Oxford (Helen Lawton Smith), Grenoble (Michel de Bernardy), Sophia-Antipolis (Christian Longhi), Munich (Rolf Sternberg and Christine Tamásy), the Dutch Randstad (Egbert Wever), Pisa, Piacenza and NE Milano (Roberta Capello and Roberto Camagni), Goteborg (Asa Lindholm Dahlstrand), Helsinki (Ilkka Kauranen and Erkko Autio), and Barcelona (Pere Escorsa, Ramon Maspons and Jaume Valls), with theoretical contributions from Edward Lorenz. The network was coordinated by David Keeble and Frank Wilkinson of the ESRC Centre for Business Research, University of Cambridge, with Clive Lawson as network rapporteur. While not all network members have contributed to this book, the chapters which follow are greatly indebted to the collective endeavours and debates of the whole network.[1]

The aim of this book is to provide a thematic overview of key findings of the network, rather than to present a series of individual regional case studies.[2] Each chapter investigates a different and specific theme identified from both the theoretical and empirical literature as being important in understanding the processes and dynamics of regional clustering and growth of high-technology SMEs, and attempts to draw general conclusions from comparison of findings across a range of the network's case study regions. The book thus begins with an introductory discussion of key contextual issues and definitions (chapter 1, Keeble and Wilkinson), followed by an examination of the nature of recent European high-technology region evolutionary trajectories and the forces driving these trajectories (chapter 2, Longhi and Keeble). Chapter 3 (Tamásy and Sternberg) outlines the nature and reviews the impact of different regional institutional and policy contexts, while chapter 4 (Lawton Smith and de Bernardy) examines the role of local universities and public research institutes in shaping the recent growth of European regional clusters of technology-based SMEs. The issue of the nature and importance of inter-SME collaborative links and networks within these regions is addressed in chapter 5 (Camagni and Capello), and that of the impact on regional clusters of large firms, especially through their activities in spinning-off and acquiring regional technology-based SMEs, is discussed in chapter 6 (Lindholm Dahlstrand). Chapter 7 (Lawson) provides a theoretical perspective on the concept of 'regional collective learning', which is increasingly being argued by commentators as being of central importance for understanding how successful regional SME clusters maintain and enhance their innovativeness in a technologically-dynamic world: while chapter 8 (Keeble) assesses the extent and importance of specific regional collective learning processes identified by the network's research as currently operating in European high-technology regions. Finally, chapter 9 (Wilkinson and Moore) considers some of the main policy implications and conclusions for promoting and supporting Europe's technology-based regional SME clusters arising from the network's findings.

1.2 High-Technology SMEs: Conceptualisation and Definition

The term 'high-technology' is widely – and loosely – used as a catch-all phrase usually denoting industries producing technologically-advanced, sophisticated and changing products. Its uncritical usage has in the past led researchers such as McArthur (1990) to argue that the concept of 'high-technology' is too 'chaotic' to justify continued use. McArthur's preferred alternative two-fold categorisation of technologically-based activities into those involving 'widely diffusing' and 'newly emerging' technologies has not however itself been adopted subsequently, while the academic literature continues to find the 'high-technology' categorisation useful (Anselin et al., 1997; Westhead and Storey, 1997; Bolland and Hofer, 1998; Garnsey, 1998; Pfirrman, 1998; Oakey and Mukhtar, 1999). In this book, the terms 'high-technology', 'technology-intensive' and 'technology-based' are used broadly and interchangeably to refer to firms and industries whose products or services embody new, innovative and advanced technologies developed by the application of scientific and technological expertise. Such firms almost invariably regard such expertise and resultant technological leadership as the firm's leading competitive advantage, and are usually identified in practice by high R&D-intensity (high levels of research and development expenditure and/or employment relative to turnover or total workforce) (Aydalot and Keeble, 1988b; Keeble, 1992).

Butchart (1987) used R&D-intensity indicators for UK sectors to identify a set of high-technology industries which can be contrasted with medium- or low-technology industries. Butchart's list of high-technology sectors (Table 1.1), though now somewhat dated, highlights the important fact that some of the most rapidly-growing technology-based sectors are service industries, such as computer software and services, rather than manufacturing industries. Computer services and software in fact recorded the largest volume of growth in numbers of businesses – overwhelmingly small businesses – of all 4-digit sectors of the British economy between 1985 and 1990 (Keeble et al., 1992), while employment in this high-technology service sector grew by no less than 186 per cent (+102 thousand jobs) between 1981 and 1994 (Keeble and Oakey, 1995). This growth has continued in the 1990s (see chapter 2, section 2.7).

Table 1.1 High-technology industries in the United Kingdom

Standard Industrial Classification (1980) Activity Heading and Industry Description

2514	Synthetic resins and plastics materials
2515	Synthetic rubber
2570	Pharmaceutical products
3301	Office machinery
3302	Electronic data processing equipment
3420	Basic electrical equipment
3441	Telegraph and telephone apparatus and equipment
3442	Electrical instruments and control systems
3443	Radio and electronic capital goods
3444	Components other than active components mainly for electronic equipment
3453	Active components and electronic sub-assemblies
3640	Aerospace equipment manufacturing and repairing
3710	Measuring, checking and precision instruments and apparatus
3720	Medical and surgical equipment and orthopaedic appliances
3732	Optical precision instruments
3733	Photographic and cinematographic equipment
7902	Telecommunications services
8394	Computing services
9400	Research and development services

Source: Butchart 1987.

This said, it is clear that many research-based firms producing technology-intensive goods and services are to be found in sectors other than those listed in Table 1.1, while conversely, some firms in these sectors are neither research-intensive nor technologically-dynamic. Indeed, recent Cambridge research by Hughes (1998), applying a firm-level approach to high-technology definition (R&D expenditure relative to sales) to data from a large national stratified random sample of British manufacturing and business service SMEs, has found that there are considerably more research-intensive firms outside these sectors (8.3 per cent of the sample) than within them (5.3 per cent). For these reasons, the approach taken to identifying technology-based SMEs in the network's investigations of the various European clusters studied has been inclusive rather than exclusive, industry lists such as Butchart's providing only a starting point for identifying local populations of high-technology SMEs which satisfy the broad definition given in the first paragraph above.

The best general definition of a small firm probably remains that of the UK Bolton Committee in its 1971 Report on Small Firms, namely that a small firm is an independent business which has only a relatively small share of its market and is managed by its owner(s) in a personalised way (Storey, 1994, p. 9). By extension, the European Commission has since 1996 defined small and medium-sized enterprises (SMEs) as having no more than 250 employees, an annual sales turnover of no more than 40 million ecus (1996), and no more than 25 per cent external ownership by larger firms. Again, the approach adopted in the network's case studies has been to focus on SMEs which broadly fit this latter definition, while explicitly acknowledging that close links with large firms are often important to their competitive success and that successful high-technology SMEs are often prime targets for acquisition as subsidiaries by larger firms (see chapter 6).

1.3   Why are High-Technology SMEs Important?

The network's focus on high-technology SMEs can be justified on various grounds. Though such firms account for only a small proportion of European SMEs,[3] empirical research has clearly shown that they differ significantly from their more conventional counterparts, in ways which indicate a significantly more favourable and longer term economic impact upon regional and national economies and labour markets. Some of these differences are discussed in Keeble and Oakey (1995) and examined in detail in section 2.7 in relation to Hughes' (1998) recent research. Suffice it to note here that high-technology SMEs are significantly more innovative, whether this is conceptualised as involving radical technological innovations (Tether, 1995, ch. 2) or simply the development or introduction of products which are new to the firm;[4] they grow more rapidly in both employment and sales; they are much less vulnerable to closure and more likely to survive over any given period (Reid and Garnsey, 1996, 1997; Storey and Tether, 1998); they more frequently serve wider national and global markets (Keeble et al., 1998), thus generating significantly greater basic income for regional and national economies; they are much more likely to engage in collaborative and co-operative arrangements with other firms and organisations, thus exerting wider multiplier effects; and they employ much higher proportions of highly-skilled and high-income professional, scientific and managerial staff, thus enriching regional labour markets, boosting regional incomes and stimulating regional entrepreneurship via researchers spinning-off new firms from existing high-technology SMEs. Technology-based SMEs also appear to be playing a key role in the rapid growth of new and dynamic sectors of Europe's economy, such as bio-technology and computer software and internet applications (Ernst and Young, 1999), and hence in structural economic change of a more fundamental kind.

1.4 The Growth of High-Technology SMEs in Europe: Nature and Causes

Although no comprehensive statistics are available, the number and importance of technology-based SMEs in Europe have undoubtedly been growing during the 1980s and 1990s (Storey and Tether, 1998; Tether and Storey, 1998).[5] A striking example is the European biotechnology sector, where numbers of firms – overwhelmingly SMEs – have more than doubled in the last four years, 1995–99 (Ernst and Young, 1995, 1999: see chapter 2, section 2.7). Another is computer software, services and applications, where numbers of enterprises, again overwhelmingly small firms employing less than 10 staff, soared in the United Kingdom by 71 per cent between 1993 and 1997 (chapter 2, section 2.7). While small firms generally undoubtedly suffered severely from the early-1990s European-wide recession (see Figure 2.1), technology-based SMEs were much less affected, with much lower closure rates and continuing growth (Reid and Garnsey, 1996, 1997).

Indeed, in the Netherlands, growth rates of R&D-intensive new firms between 1994 and 1996 were three times higher than the average for all new firms (Snijders and van Elk, 1998). Finally, Eurostat data reveal that while European economy-wide new firm creations exceeded closures in 1994-95, resulting in an overall net increase in Europe's small firm population after net losses during the early 1990s, no less than four of the six sectors singled out as recording exceptional net growth in numbers of small firms were high-technology sectors, namely computer manufacturing, computer and related services, research and development services, and telecommunications (European Commission, 1998, pp. 67–8).

The exceptional growth rates of European high-technology SMEs of course further highlight the significance of studying the evolution and functional dynamics of the regional clusters discussed in this book. The causes of rapid growth in numbers of small technology-based firms would seem primarily to lie in rapid, radical and increasingly-pervasive technological change, which is in turn fuelling buoyant growth in demand for technology-intensive products and services. Technological change and surging market demand are being spearheaded by remarkable developments in computer technology, software and applications, and their pervasive adoption in all areas of economic and social activity. But other radical technological changes based on scientific research are also involved, as with biotechnology, life sciences and materials science.

In many of these sectors, rapid demand growth is associated with the proliferation of a host of new and specialised market niches which small technology-based firms set up by scientists, engineers and professionals are particularly able to exploit. Increasing market segmentation and market niche proliferation appears to be a general trend underlying small firm growth in Europe in the 1980s and 1990s (Brusco, 1982; SBRC, 1992, pp. 18–19), in line with so-called flexible specialisation theory (Keeble, 1990): but it does appear to be of especial importance in the recent growth of technology-based SMEs (see chapter 2, section 2.7). New market opportunities for technology-based SMEs have also been opened up by institutional and policy changes, as for example with the privatisation and de-monopolisation of telecommunications in several European countries. On the supply side, low barriers to new firm entry, in the form of low capital costs and the possession by 'boffin' entrepreneurs of unique technological knowledge, are a further important contributing influence on the growth of small technology-based firms. Government policies have also sometimes helped in a modest way, through the provision of technology grants such as the UK's SMART scheme (Moore, 1993), the encouragement of private venture capital funds (Bank of England, 1996), or the development of science and technology parks and incubators (Hauschildt and Steinkuhler, 1994; Sternberg et al., 1997).

1.5 The Growth of Technology-Based SMEs in Europe: Spatial Dispersion or Regional Clustering?

This book examines in detail the growth of distinctive European regional clusters of high-technology enterprises, most of which have come into being as new and notable features of the economic landscape of Europe since the 1960s. These regional clusters parallel, albeit on a much more modest scale, the globally-famous high-technology regional complexes of the USA such as Silicon Valley and Orange County in California, Boston's Route 128, or Austin, Texas: and the full list of European technology-based SME clusters of course extends well beyond those studied here, to embrace examples such as Oulu in northern Finland (Asheim and Cooke, 1999, p. 167), Linköping[6] in southern Sweden (Jones-Evans and Klofsten, 1997), and the Western Crescent cluster focussed on Berkshire to the west of London (Hall et al., 1987). Clustering appears to be associated with exceptional rates of both technological innovation (Baptista and Swann, 1998) and new firm spin-off (Oakey, 1995, p. 13).

This said, it should be noted that in some northern European countries at least, the creation and growth of small technology-intensive firms has also taken place in a spatially dispersed pattern, involving a variety of accessible rural areas and small towns, and is not confined to regional clusters alone (Keeble, 1994, pp. 204–7; Oakey and Cooper, 1989). This spatial dispersion of technology-based industry represents part of a wider 'urban-rural shift' of economic activity which has been going on ever since the 1970s (Keeble, Owens and Thompson, 1983; Keeble, 1993; Keeble, 1999), and which seems to reflect the impact of 'enterprising behaviour' in the form of enterprise creation by entrepreneurs, professionals and technologically-qualified individuals who have earlier moved from big cities to smaller towns and accessible villages for reasons of environmental preference and higher perceived quality of life for themselves and their families (Keeble and Tyler, 1995). It is thus broadly confined to those smaller towns and rural regions whose landscapes, architecture, climate and better accessibility offer attractive living environments to urbanised individuals, such as

Östergötland County around Linköping, or rural Dorset, Wiltshire, Somerset and Devon in England, rather than to Europe's more remote, physically unattractive or climatically extreme rural areas.

1.6 Understanding Regional Clustering: Innovative Milieux, Learning Regions, and Regional Collective Learning

Since the 1970s, an enormous literature has developed dealing with one or other aspect of the growth of regional clusters of high-technology industry in North America and western Europe. The research reported in this book has focussed its attention upon three overlapping issues arising from this literature which appear to be central to the successful development and continuing growth of the regional clusters studied. These are the vital importance of firm innovation, of inter-firm and organisation networking, and of regional collective learning within territorially-localised socio-cultural and institutional environments.

The vital importance for high-technology SME competitiveness and growth of continuing innovation, especially in products and services, is powerfully attested by numerous empirical and theoretical studies (see for example chapter 2, section 2.7). Perhaps less widely understood however is the importance for successful innovation by such SMEs of networking, collaborative and co-operative linkages with other firms and institutions such as knowledge centres (universities, public research laboratories), particularly within a particular territorial context in which localised interaction is facilitated. A 'business network' may be defined as 'an integrated and co-ordinated set of ongoing economic and non-economic relations embedded within, among and outside business firms' (Yeung, 1994, p. 476). Local networking represents a key mechanism (see chapter 8) whereby particular regional clusters may over time develop a collective learning capacity for creating, diffusing and elaborating new technological and organisational knowledge within and between the cluster's constituent firms. The development of such a localised collective learning capability appears to be a fundamental common characteristic of all the successful regional clusters studied by this European research network. This empirical finding echoes the recent theoretical conclusion drawn by Maskell and Malmberg (1999a, p. 180) more generally that 'the path-dependent and interactive character of knowledge creation is a key to understanding the contemporary emergence and reproduction of spatial agglomerations of related firms'. Indeed, the whole special issue of the Cambridge Journal of Economics in which this paper appears (Amin and Wilkinson, 1999) bears witness to the judgment that learning and knowledge creation are both crucially important for developing a firm's core competences, and invariably reflect collective interaction extending beyond the boundaries of the firm itself.

In its discussions and the chapters which follow, the research network has also frequently made use of the theoretical concept of the 'innovative milieu' developed since the early 1980s by members of the GREMI, several of whom were also members of the network.[7] The innovative milieu concept focuses on innovation by firms as the key 'motor of growth' (Bramanti and Ratti, 1997, p. 5) of local and national economies, and argues that 'innovation is fundamentally a collective process', 'innovation is a complex and interactive process', 'innovation stems from a creative combination of generic know-how and specific competencies', and 'territorial organisation is an essential component of the process of techno-economic creation' (Bramanti and Ratti, 1997, p. 5). Given this innovation focus, Camagni (1991b, p. 130) defines an innovative milieu or local industrial environment which promotes innovation 'as a set of territorial relationships encompassing in a coherent way a production system, different economic and social actors, a specific culture and representation system, and generating a dynamic collective learning process' whereby innovation is sustained and uncertainty minimised. The innovative milieu concept thus emphasises the importance for firm innovative activity of inter-firm relationships, territorial socio-economic embeddedness, and dynamic local collective learning processes, all of which resonate closely with the network's own empirical findings.

A further closely related concept to the innovative milieu, with its stress on localised collective learning, is that of 'the learning region'. This concept, which is attracting considerable attention from academics and policy makers (Florida, 1995; Asheim, 1996; Macleod, 1996; Morgan, 1997; Simmie, 1997; Hudson, 1999; Maskell and Malmberg, 1999b; Kirat and Lung, 1999), is derived by logical extension from earlier conceptualisations of innovative national economies as 'learning economies' (Lundvall and Johnson, 1994; Gregersen and Johnson, 1997; Lundvall and Borras, 1998).[8] And both the learning economy and learning region literature overlap and are closely enmeshed with recent literature on 'national innovation systems' (Lundvall, 1992; Freeman, 1995; Edquist, 1997) and 'regional innovation systems' (Braczyk, Cooke and Heidenreich, 1998; Cooke, 1998; Cooke and

Morgan, 1998, ch. 3; Cooke, Uranga and Etxebarria, 1998; Cooke, Boekholt and Todtling, 1998; Asheim and Cooke, 1999). The learning region – and regional innovation system – literature argues that in modern capitalist economies, 'knowledge is the most strategic resource and learning the most important process' (Lundvall 1994) of economic development, that successful innovative regions can be viewed as 'externalized learning institutions' (Cooke and Morgan, 1998, p. 66), and that a learning region of this kind is characterised by 'a particular structured combination of institutions strategically focused on technological support, learning and economic development' (Pratt, 1997, 128). Put slightly differently, from an innovation system perspective, 'regions which possess the full panoply of innovation organizations set in an institutional milieu where systemic linkage and interactive communication among the innovation actors is normal, approach the designation of a regional innovation system.' ... 'These are (regional) systems which combine learning with upstream and downstream innovation capability' (Cooke and Morgan, 1998, p. 71).

The learning region and regional innovation systems literature thus focus attention primarily on the nature and role of regional institutions and organisations which may facilitate knowledge development and learning by local firms. In contrast, and while sharing many insights with this literature, the 'regional collective learning' perspective which is the preferred conceptual framework adopted in this book, focuses more on processes of regional inter-firm networking and interaction which have evolved spontaneously between firms themselves, while clearly acknowledging that such interaction is embedded in wider regional institutional and socio-cultural relationships. The concept of regional collective learning is discussed in some detail in Lawson and Lorenz (1999), and in chapters 7 and 8 of this book. But a simple definition is that regional collective learning involves 'the creation and further development of a base of common or shared knowledge among individuals making up a productive system which allows them to co-ordinate their actions in the resolution of the technological and organisational problems they confront' (Lorenz, 1996). The creation and development of such a localised and interactive knowledge base can involve both 'conscious' and 'unconscious' mechanisms, an example of the former being deliberate research collaboration between local SMEs or between an SME and a local university, examples of the latter being the spontaneous movement of 'embodied expertise' and knowhow in the form of researchers, managers and skilled workers within the regional labour market and via entrepreneur spin-off from existing local firms or organisations to create new technology-intensive firms. Both conscious and unconscious mechanisms involving interaction and diffusion of knowledge between firms and organisations may thus generate a regional collective learning capability which sustains continuing innovation by the cluster's firms.

1.7 Regional SME Networks, Large Firms, and Universities

The recent growth of the European regional high-technology SME clusters studied by the network owes much, though to different degrees in different regions, to the role of inter-SME linkages and networks, links with large multinational firms, and links with knowledge centres such as universities and public research institutes. The impact of these different types of links is examined in detail in chapters 4, 5 and 6. But three brief points warrant emphasis in this introductory review.

First, as Storper (1995) has emphasised, high-technology SME clusters often exhibit a relatively limited degree of internalised traditional input-output linkages, in the form of supplier-customer relationships, since their markets and customers are usually national and indeed global (see section 1.8). This is supported by the findings of the European network,[9] which instead, as with Storper's research, bear witness to the importance of links between regional SMEs in the form of 'untraded interdependencies'. Such interdependencies embrace informal as well as formal information and collaboration networks, interactions through local labour markets, and shared cultures, conventions and rules for developing relationships and interpreting and exchanging knowledge. The network's research into the Cambridge high-technology cluster, for example, thus reveals that while local SMEs rank proximity to local customers only thirteenth out of 19 region-specific advantages (and proximity to suppliers tenth), 'informal local access to innovative people, ideas and technologies' was ranked third most important (Keeble et al., 1999).

Second, the network's research (see chapter 2) supports the view proposed by a number of recent commentators (de Vet, 1993; Veltz, 1993; Santangelo, 1999) that in contrast to earlier decades, large multinational firms are in the 1990s increasingly being attracted to, and locating R&D activity in, Europe's regional technology-based SME clusters. This reflects the growing primacy and imperative of continuous technological and product innovation if large firms are to maintain their market share in an intensely competitive global market-place, and the resultant importance of access to the regionally-specific and embedded competences and innovative ideas generated by local small firms and knowledge centres. This suggests that large firms, with their global communication channels and market reach, may be becoming more important actors in the evolution and growth of Europe's high-technology clusters, notwithstanding high recent rates of new enterprise creation in many cases.

Third, though almost certainly growing in importance during the 1990s, the role of regional universities in stimulating the growth of regional SME technology clusters appears to operate less through formal research collaboration and technology transfer schemes than through a range of informal and associational relationships. As with inter-firm links, therefore, measurement only of frequencies of formal collaborative links between local SMEs and universities may seriously undervalue the wider and very important role played by many such knowledge centres in shaping and enabling cluster growth. This wider role is illustrated, for example, by the longterm significance of university and research institute spin-offs in several clusters (Goteborg, Cambridge, Linköping, Sophi-Antipolis). Again, this argument is supported by evidence from Cambridge, where the network's research shows that formal research links with Cambridge university were ranked only eleventh in the list of 19 region-specific development advantages, but that the 'credibility, reputation and prestige of a Cambridge address' created by the presence of the university was ranked no less than second, its importance being reported by 70 per cent of the high-technology SMEs surveyed (Keeble et al., 1999).

1.8 Europe's High-Technology Clusters in a Globalising World Economy

The final issue which warrants emphasis in this introduction is the great importance in the growth and competitive success of the European high-technology SME clusters studied of wider national and global innovation networks, collaborative links and market relationships. In a globalising world (Amin and Thrift, 1994; Daniels and Lever, 1996; Dicken, 1998) where technological change pays no respect to national let alone regional boundaries, geographical clustering and its numerous benefits to firms in the cluster are inadequate on their own to ensure continuing successful innovation and firm growth. Equally important are global and national links and networks.

This view is clearly argued by Camagni (1991b, pp. 134–41), who points out that in a technologically-dynamic and highly uncertain world, local 'innovative milieu' advantages undoubtedly have their limits, and must be seen in conjunction with the parallel importance of wider inter-firm networks as an essential means of access to information on rapidly-changing technologies and market opportunities. This is particularly true 'in those areas of production characterised by fast innovation and technological change' (Camagni, 1991b, p. 137). Indeed, Camagni (1991b, p. 139) argues explicitly that in such sectors, local firm involvement in wider national and global networks is absolutely essential for long-term regional growth, and that 'the "milieu" has to open up to external energy in order to avoid "entropic death" and a decline in its own innovative capacity'. For Camagni (1991b, p. 139), regional collective learning or 'milieu' relationships and [wider] network relationships appear as complementary and mutually reinforcing "operators", the former linking the firm to its contiguous environment through mainly informal, tacit (and often even overlooked and apparently unappreciated) relationships, the latter linking it explicitly to selected partners in its [wider] operational environment'.

These judgments are powerfully supported by the empirical findings of the European network. Many SMEs in the European clusters studied possess close functional links with firms and knowledge centres elsewhere in their own country and abroad, and view such wider networks as very important for successful research and innovation (Keeble et al., 1999). Most clusters are thus very open to global influences, notwithstanding the parallel importance of regionally-embedded processes and networks. Indeed, the network's research into the Cambridge and Oxford technology-based clusters reveals that the most globalised of the high-technology SMEs surveyed also tend to be the most locally-embedded and networked, a finding which suggests that successful globalisation by high-technology SMEs may actually be enabled by local networking and research and technology collaboration (Keeble et al., 1998). Regional and global networking thus appear to be complementary, not alternative, processes whereby high-technology SMEs sustain their innovative activity and competitive advantage.

Notes

1   During the course of its work, the network has produced four substantive reports on the

development of these European regional clusters of high-technology SMEs. These cover the themes of regional institutional and policy frameworks; university research links and spin-offs; networks, links and large firm impacts; and collective learning processes and knowledge development (Keeble and Lawson, 1996, 1997a, 1997b, 1998). A Final Report on the network's findings was submitted to the European Commission in March 1999 (Keeble and Wilkinson, 1999a). We gratefully acknowledge the support and encouragement of Virginia Vitorino, DG XII liaison officer for the network, and the cheerful efficiency of Anita Biggs and Linda Brosnan of the ESRC Centre for Business Research, University of Cambridge, successive network secretaries. We are also indebted to Ian Agnew of Cambridge University's Department of Geography for drawing most of the maps and diagrams in this book.

2. A series of regional case studies by members of the network were published in a Special Issue of the journal Regional Studies in June 1999 (Volume 33, Number 4, pp. 295–400), edited by Keeble and Wilkinson (1999b).

3. For example, only about 14 per cent of the UK sample of manufacturing and business service SMEs surveyed by the Cambridge University ESRC Centre for Business Research in 1997 were classed as high-technology firms using either the Butchart or individual firm R&D-intensity criteria. The proportion in the whole UK SME population would thus be much smaller than this.

4. It is noteworthy that the much higher innovation rates recorded by high-technology SMEs are not due to the coincidental fact that such firms are on average younger and smaller than their non-high technology counterparts: multiple regression analysis of innovation propensity against age, size and high-technology status in combination reveals that the only one of these three variables to record a statistically significant, and highly positive, relationship is high-technology status (Keeble and Oakey, 1995, p. 11).

5. These authors point out that while numbers of small European high-technology firms grew rapidly during the 1980s, the period of their study, in both services and manufacturing, employment in high-technology manufacturing as a whole actually fell. They ascribe this to large firm down-sizing, but although not mentioned in their article, it undoubtedly also reflects the early 1990s recession, since their national data sets often terminate in 1992–94. This must bias their results. Employment and firm growth in high-technology services was very rapid in both cases.

6. By 1999, the Linköping cluster contained around 250 firms employing over 18,000 workers: the regional rate of new firm formation has intensified during the 1990s compared with the later 1980s, with at least 40 new technology-based firms created 1997–98 (www.smil.se).

7. Notably Roberto Camagni, President of GREMI, Roberta Capello, Michel de Bernardy and Christian Longhi. GREMI stands for Groupe de Recherche Européen sur les Milieux Innovateurs.

8. It is also closely related to the work of Michael Storper (1995, 1997), with its powerfully argued case that 'the answer to the principal dilemma of contemporary economic geography – the resurgence of regional economies [lies partly in] the association of organizational and technological learning with agglomeration' and that 'the region is a key, necessary element in the "supply architecture" for learning and innovation' (Storper, 1995, p. 210). Storper's stress on the importance for innovation and learning of regionally-embedded 'untraded interdependencies' and common regional socio-cultural conventions, rules and norms of behaviour has also influenced, and is supported by the findings of, the European SME network.

9 An exception here may be the Italian high-technology milieux studied by Capello (1999), which exhibit high levels of local customer-supplier purchases in accordance with the traditional Italian industrial district model.

References

Amin, A. and Thrift, N. (eds) (1994), *Globalization, Institutions and Regional Development in Europe*, Oxford University Press, Oxford.
Amin, A. and Wilkinson, F. (eds) (1999), 'Learning, Proximity and Industrial Performance', Special Issue of *Cambridge Journal of Economics*, vol. 23, pp. 121–260.
Anselin, L., Varga, A. and Acs, Z. (1997), 'Local Geographic Spillovers Between University Research and High Technology Innovations', *Journal of Urban Economics*, vol. 42, pp. 422–48.
Asheim, B.T. (1996), 'Industrial Districts as 'Learning Regions': A Condition for Prosperity?', *European Planning Studies*, vol. 4, pp. 379–400.
Asheim, B.T. and Cooke, P. (1999), 'Local Learning and Interactive Innovation Networks in a Global Economy', in E.J. Malecki and P. Oinas (eds), *Making Connections: Technological Learning and Regional Economic Change*, Ashgate, Brookfield USA, pp. 145–78.
Aydalot, P. (ed.) (1986), *Milieux Innovateurs en Europe*, GREMI, Paris.
Aydalot, P. and Keeble, D. (eds) (1988a), *High Technology Industry and Innovative Environments: the European Experience*, Routledge, London.
Aydalot, P. and Keeble, D. (1988b), 'High-Technology Industry and Innovative Environments in Europe: An Overview', in P. Aydalot and D. Keeble (eds), *High Technology Industry and Innovative Environments: the European Experience*, Routledge, London, pp. 1–21.
Bank of England (1996), *The Financing of Technology-Based Small Firms*, Bank of England, London.
Baptista, R. and Swann, P. (1998), 'Do Firms in Clusters Innovate More?', *Research Policy*, vol. 27, pp. 525–40.
Bolland, E.J. and Hofer, C.W. (1998), *Future Firms: How America's High Technology Companies Work*, Oxford University Press, New York.
Braczyk, H.-J., Cooke, P. and Heidenreich, M. (eds) (1998), *Regional Innovation Systems: The Role of Governances in a Globalized World*, UCL Press, London.
Bramanti, A. and Ratti, R. (1997), 'The Multi-Faceted Dimensions of Local Development', in Ratti, R., Bramanti, A. and Gordon, R. (eds), *The Dynamics of Innovative Regions: The GREMI Approach*, Ashgate, Aldershot, pp. 3–44.
Brusco, S. (1982), 'The Emilian Model: Productive Decentralisation and Social Integration', *Cambridge Journal of Economics*, vol. 6, pp. 167–84.
Butchart, R.L. (1987), 'A New UK Definition of the High Technology Industries', *Economic Trends*, no. 400, pp. 82–8.
Camagni, R. (ed.) (1991a), *Innovation Networks: Spatial Perspectives*, Belhaven Press, London.
Camagni, R. (1991b), 'Local 'Milieu', Uncertainty and Innovation Networks: Towards a New Dynamic Theory of Economic Space', in R. Camagni (ed.), *Innovation Networks: Spatial Perspectives*, Belhaven Press, London, pp. 121–43.

Capello, R. (1999), 'Spatial Transfer of Knowledge in High Technology Milieux: Learning Versus Collective Learning Processes', *Regional Studies*, vol. 33, pp. 353–65.
Cooke, P. (1998), 'Introduction: Origins of the Concept', in Braczyk, H.-J., Cooke, P. and Heidenreich, M. (eds), *Regional Innovation Systems: The Role of Governances in a Globalized World*, UCL Press, London, pp. 2–25.
Cooke, P., Boekholt, P. and Todtling, F. (1998), *Regional Innovation Systems: Designing for the Future*, Final Report to DG12 of the REGIS TSER project, Centre for Advanced Studies in the Social Sciences, University of Wales, Cardiff.
Cooke, P. and Morgan, K. (1998), *The Associational Economy: Firms, Regions and Innovation*, Oxford University Press, Oxford.
Cooke, P., Uranga, M. G., and Etxebarria, G. (1988), 'Regional Systems of Innovation: An Evolutionary Perspective', *Environment and Planning A*, 30, 1563–84.
Daniels, P.W. and Lever, W. F. (eds) (1996), *The Global Economy in Transition*, Addison Wesley Longman, Harlow.
de Vet, J. (1993), 'Globalisation and Local and Regional Competitiveness', *STI Review*, vol. 13, pp. 89–121.
Dicken, P. (1998), *Global Shift: Transforming the World Economy*, 3rd edn, Paul Chapman Publishing, London.
Edquist, C. (ed.) (1997), *Systems of Innovation: Technologies, Institutions and Organizations*, Pinter, London.
Ernst and Young (1995), *European Biotech 95: Gathering Momentum*, Ernst and Young International, London.
Ernst and Young (1999), *Ernst and Young's European Life Sciences 99: Sixth Annual Report, Communicating Value*, Ernst and Young International, London.
European Commission (1998), *Enterprises in Europe: Fifth Report*, Office for Official Publications of the European Communities, Luxembourg.
Florida, R. (1995), 'Toward the Learning Region', *Futures*, vol. 27, pp. 527–36.
Freeman, C. (1995), 'The 'National System of Innovation' in Historical Perspective', *Cambridge Journal of Economics*, vol. 19, pp. 5–24.
Garnsey, E. (1998), 'The Genesis of the High-Technology Milieu: A Study in Complexity', *International Journal of Urban and Regional Research*, vol. 22, pp. 361–77.
Gregersen, B. and Johnson, B. (1997), 'Learning Economies, Innovation Systems and European Integration', *Regional Studies*, vol. 31, pp. 479–90.
Hall, P., Breheny, M., McQuaid, R. and Hart, D. (1987), *Western Sunrise: The Genesis and Growth of Britain's Major High Tech Corridor*, Allen and Unwin, London.
Hauschildt, J. and Steinkuhler, R.H. (1994), 'The Role of Science and Technology Parks in NTBF Development', in R. Oakey (ed.), *New Technology-Based Firms in the 1990s*, Paul Chapman, London, pp. 181–91.
Hudson, R. (1999), '"The Learning Economy, the Learning Firm and the Learning Region": a Sympathetic Critique of the Limits of Learning', *European Urban and Regional Studies*, vol. 6, pp. 59-72.
Hughes A. (1998), 'High-Tech Firms and High-Tech Industries: Finance, Innovation and Human Capital', paper presented at Conference on *SMEs and Innovation Policy: Networks, Collaboration and Institutional Design*, ESRC Centre for Business Research, University of Cambridge, Cambridge.
Jones-Evans, D. and Klofsten, M. (1997), 'Universities and Local Economic Development: The Case of Linköping', *European Planning Studies*, vol. 5, pp. 77–93.

Keeble, D. (1990), 'Small Firms, New Firms and Uneven Regional Development in the United Kingdom', *Area*, vol. 22, pp. 342–45.
Keeble, D. (1992), 'High Technology Industry and the Restructuring of the UK Space Economy', in P. Townroe and R. Martin (eds), *Regional Development in the 1990s: The British Isles in Transition*, Jessica Kingsley, London, pp. 172–81.
Keeble, D. (1993), 'Small Firm Creation, Innovation and Growth and the Urban-Rural Shift', in Curran, J. and Storey, D. (eds), *Small Firms in Urban and Rural Locations*, Routledge, London, pp. 54–78.
Keeble, D. (1994), 'Regional Influences and Policy in New Technology-Based Firm Creation and Growth', in R. Oakey (ed.), *New Technology-Based Firms in the 1990s*, Paul Chapman, London, pp. 204–18.
Keeble, D. (1999), 'Urban Regeneration, SMEs and the Urban-Rural Shift in the United Kingdom', in E. Wever (ed.), *Cities in Perspective: Part 1, Economy, Planning and the Environment*, Urban Research Centre Utrecht, University of Utrecht, ch. 3.
Keeble, D., Bryson, J. and Wood, P. (1992), 'The Rise and Role of Small Service Firms in the United Kingdom', *International Small Business Journal*, vol. 11, pp. 11–22.
Keeble, D. and Lawson, C. (eds) (1996), *Regional Institutional and Policy Frameworks for High-Technology SMEs in Europe*, Report on Presentations and Discussions, Sophia-Antipolis Meeting of the TSER European Network on 'Networks, Collective Learning and RTD in Regionally-Clustered High-Technology SMEs', ESRC Centre for Business Research, University of Cambridge.
Keeble, D. and Lawson, C. (eds) (1997a), *University Research Links and Spin-Offs in the Evolution of Regional Clusters of High-Technology SMEs in Europe*, Report on Presentations and Discussions, Munich Meeting of the TSER European Network on 'Networks, Collective Learning and RTD in Regionally-Clustered High-Technology SMEs', ESRC Centre for Business Research, University of Cambridge.
Keeble, D. and Lawson, C. (eds) (1997b), *Networks, Links and Large Firm Impacts on the Evolution of Regional Clusters of High-Technology SMEs in Europe*, Report on Presentations and Discussions, Barcelona Meeting of the TSER European Network on 'Networks, Collective Learning and RTD in Regionally-Clustered High-Technology SMEs', ESRC Centre for Business Research, University of Cambridge.
Keeble, D. and Lawson, C. (eds) (1998), *Collective Learning Processes and Knowledge Development in the Evolution of Regional Clusters of High-Technology SMEs in Europe*, Report on Presentations and Discussions, Goteborg Meeting of the TSER European Network on 'Networks, Collective Learning and RTD in Regionally-Clustered High-Technology SMEs', ESRC Centre for Business Research, University of Cambridge.
Keeble, D., Lawson, C., Lawton Smith, H., Moore, B. and Wilkinson, F. (1998), 'Internationalisation Processes, Networking and Local Embeddedness in Technology-Intensive Small Firms', *Small Business Economics*, vol. 11, pp. 327–42.
Keeble, D., Lawson, C., Moore, B. and Wilkinson, F. (1999), 'Collective Learning Processes, Networking and 'Institutional Thickness' in the Cambridge Region', *Regional Studies*, vol. 33, pp. 319–32.
Keeble, D. and Oakey, R. (1995), 'Spatial Variations in Innovation in High-Technology Small and Medium-Sized Enterprises: A Review', ESRC Centre for Business Research, University of Cambridge, mimeo.
Keeble, D., Owens, P.L. and Thompson, C. (1983), 'The Urban Rural Manufacturing Shift in the European Community', *Urban Studies*, vol. 20, pp. 405–18.

Keeble, D. and Tyler, P. (1995), 'Enterprising Behaviour and the Urban-Rural Shift', *Urban Studies*, vol. 32, pp. 975–97.

Keeble, D. and Wilkinson, F. (eds) (1999a), *Networking and Collective Learning in Regionally-Clustered High-Technology SMEs in Europe*, Final Report to DG XII – TSER, European Commission, ESRC Centre for Business Research, University of Cambridge.

Keeble, D. and Wilkinson, F. (1999b), 'Collective Learning and Knowledge Development in the Evolution of Regional Clusters of High Technology SMEs in Europe', *Regional Studies*, vol. 33, pp. 295–303.

Kirat, T. and Lung, Y. (1999), 'Innovation and Proximity: Territories as Loci of Collective Learning Processes', *European Urban and Regional Studies*, vol. 6, pp. 27–38.

Lawson, C. and Lorenz, E. (1999), 'Collective Learning, Tacit Knowledge and Regional Innovative Capacity', *Regional Studies*, vol. 33, pp. 305–17.

Lorenz, E. (1996), 'Collective Learning Processes and the Regional Labour Market', unpublished research note, European Network on Networks, Collective Learning and RTD in Regionally-Clustered High-Technology SMEs.

Lundvall, B.-A. (ed.) (1992), *National Systems of Innovation: Towards a Theory of Innovation and Interactive Learning*, Pinter, London.

Lundvall, B.-A. (1994), 'The Learning Economy: Challenges to Economic Theory and Policy', paper to the EAEPE Conference, Copenhagen, 27–9 Oct.

Lundvall, B.-A. and Borrás, S. (1998), *The Globalising Learning Economy: Implications for Innovation Policy*, European Commission, DG XII, Luxembourg.

Lundvall, B.-A. and Johnson, B. (1994), 'The Learning Economy', *Journal of Industry Studies*, vol. 1, pp. 23–41.

Macleod, G. (1996), 'The Cult of Enterprise in a Networked, Learning Region? Governing Business and Skills in Lowland Scotland', *Regional Studies*, vol. 30, pp. 749–55.

Maskell, P. and Malmberg, A. (1999a), 'Localised Learning and Industrial Competitiveness', *Cambridge Journal of Economics*, vol. 23, pp. 167–85.

Maskell, P. and Malmberg, A. (1999b), 'The Competitiveness of Firms and Regions: 'Ubiquitification' and the Importance of Localised Learning', *European Urban and Regional Studies*, vol. 6, pp. 9–25.

McArthur, R. (1990), 'Replacing the Concept of High Technology: Towards a Diffusion-Based Approach', *Environment and Planning A*, vol. 22, pp. 811–28.

Moore, I. (1993), 'Government Finance for Innovation in Small Firms: the Impact of SMART', in M. Dodgson and R. Rothwell (eds), *Small Firms and Innovation: The External Influences*, special publication of the *Journal of Technology Management*, pp. 104–18.

Morgan, K. (1997), 'The Learning Region: Institutions, Innovation and Regional Renewal', *Regional Studies*, vol. 31, pp. 491–503.

Oakey, R.P. (1995), *High-Technology New Firms: Variable Barriers to Growth*, Paul Chapman, London.

Oakey, R.P. and Cooper, S.Y. (1989), 'High Technology Industry, Agglomeration and the Potential for Peripherally Sited Small Firms', *Regional Studies*, vol. 23, pp. 347–60.

Oakey, R.P. and Mukhtar, S.-M. (1999), 'UK High-Technology Small Firms in Theory and Practice: a Review of Recent Trends', *International Small Business Journal*, vol. 17, pp. 48–64.

Pfirrman, O. (1998), 'Small Firms in High Tech: A European Analysis', *Small Business Economics*, vol. 10, pp. 227–41.

Pratt, A. (1997), 'The Emerging Shape and Form of Innovation Networks and Institutions', in Simmie, J. (ed.), *Innovation, Networks and Learning Regions?*, Jessica Kingsley, London, pp. 124–36.

Ratti, R., Bramanti, A. and Gordon, R. (eds) (1997), *The Dynamics of Innovative Regions: The GREMI Approach*, Ashgate, Aldershot.

Reid, S. and Garnsey, E. (1996), 'High-Technology – High Risk? The Myth of the Fragile Firm', *University of Cambridge Research Papers in Management Studies*, 1996, no. 10.

Reid, S. and Garnsey, E. (1997), 'High-Tech – High Risk?', *New Economy*, vol. 4, pp. 131–5.

Santangelo, G.D. (1999), 'Inter-European Regional Dispersion of Corporate Research Activity in Information and Communication Technology: The Case of German, Italian and UK Regions', *University of Reading, Department of Economics, Discussion Papers in International Investment and Management*, Series B, no. 268.

SBRC (Small Business Research Centre) (1992), *The State of British Enterprise*, Small Business Research Centre, University of Cambridge, Cambridge.

Scott, A. (1988), *New Industrial Spaces: Flexible Production Organisation and Regional Development in North America and Western Europe*, Pion, London.

Simmie, J. (ed.) (1997), *Innovation, Networks and Learning Regions?* Jessica Kingsley, London.

Sternberg, R., Behrendt, H., Seeger, H. and Tamasy, C. (1997), *Bilanz eines Booms – Wirkungsanalyse von Technologie- und Grunderzentren in Deutschland*, Dortmund.

Storey, D.J. (1994), *Understanding the Small Business Sector*, Routledge, London.

Storey, D.J. and Tether, B.S. (1998), 'New Technology-based Firms in the European Union: An Introduction', *Research Policy*, vol. 26, pp. 933–46.

Storper, M. (1993), 'Regional 'Worlds' of Production: Learning and Innovation in the Technology Districts of France, Italy and the USA', *Regional Studies*, vol. 27, pp. 433–55.

Storper, M. (1995), 'The Resurgence of Regional Economies, Ten Years Later: The Region as a Nexus of Untraded Interdependencies', *European Urban and Regional Studies*, vol. 2, pp. 191–221.

Storper, M. (1997), *The Regional World: Territorial Development in a Global Economy*, Guilford Press, New York.

Tether, B.S. (1995), *Virtual Panacea and Actual Reality: Small Firms, Innovation and Employment Creation; Evidence from Britain during the 1980s*, unpub. DPhil thesis, University of Sussex, Brighton.

Tether, B.S. and Storey, D.J. (1998), 'Small Firms and Europe's High Technology Sectors: A Framework for Analysis and Some Statistical Evidence', *Research Policy*, vol. 26, pp. 947–71.

Veltz, P. (1993), 'D'une Géographie des Coûts et une Géographie de l'Organisation: Quelques Thèses sur l'Evolution des Rapports Entreprises/Territoires', *Revue Economique*, vol. 4, pp. 671–84.

Westhead, P. and Storey, D. (1997), 'Financial Constraints on the Growth of High Technology Small Firms in the United Kingdom', *Applied Financial Economics*, vol. 7, pp. 197–202.

Yeung, H.W. (1994), 'Critical Reviews of Geographical Perspectives on Business Organisations and the Organization of Production: Towards a Network Approach', *Progress in Human Geography*, vol. 18, pp. 460–90.

2 High-Technology Clusters and Evolutionary Trends in the 1990s

CHRISTIAN LONGHI AND DAVID KEEBLE

The different regional clusters of innovative high-technology SMEs whose evolution is reviewed in this chapter, and more generally throughout this book, play a distinctive role in the European economic and technological scene. As far as the creation of wealth is concerned, regions matter; and SMEs often today have a pivotal role in the development process. But this role is mediated through very diverse local institutional arrangements, origins and structures. Some old traditional industrial regions have recently been able to regenerate, while new regions have emerged. Europe's current economic landscape results from major changes in the economic environment confronting European countries since the 1970s. This chapter will review the major evolutionary trends characterising the development of the SME regional clusters since their formation, and their further adaptation in the 1990s. To this end, the chapter begins by examining the changing industrial structure of advanced economies since the 1970s, chiefly in relation to the increased importance of SMEs in economic and innovative processes. It then focusses on parallel changes in the location and organisational strategies of large firms, especially regarding the innovative process. Finally, after reviewing the differing origins and characteristics of each of the regional clusters of high-technology SMEs studied by the Network, it will assess the nature and causes of evolutionary trends in the 1990s.

2.1 The Evolving Importance of SMEs

Several important studies (Acs and Audretsch, 1993; Audretsch, 1995; Birch, 1981; Keeble and Wever, 1986, among others) have demonstrated that since the 1970s the share of small firms in economic activity in

European and North American countries has increased significantly compared to the share of large firms. Birch (1981) reported for example that whatever else they were doing, large firms were no longer the major generators of jobs in the United States, but that most new jobs emanated from small and medium-sized firms. Between 1976 and 1988 small US firms (less than 20 workers), though providing less than 20 per cent of total employment, accounted for 37 per cent of new US jobs (Audretsch, 1995). In the UK, SMEs employing less than 100 workers accounted for 68 per cent of net job growth during the boom years 1987–89, and 126 per cent of net job growth during recession, 1989–91 (European Network for SME Research, 1994: Keeble and Bryson, 1996).

This evolutionary shift in firm size structure might be attributed to the obvious long term sectoral evolution of advanced economies from manufacturing to services. But the trend towards smaller enterprises is in fact characteristic of manufacturing as well. Thus Audretsch (1995) reports that the share of manufacturing employment accounted for by SMEs (less than 500 employees) in most developed countries increased between the 1970s and the 1980s. While the magnitude of the trend differs across economies, the direction does not; it ranges from an increase of 1.9 percentage points in the employment share of SMEs in the US between 1976 and 1986, to an increase of 10.9 percentage points in the North of Italy (7.0 percentage points in Southern Italy) between 1981 and 1987, and an increase of 9.8 percentage points between 1976 and 1986 in the UK.

In fact, as emphasised in Dosi and Salvatore (1992), the seminal contributions of Simon and Bonini (1958) and Ijiri and Simon (1977) have demonstrated a puzzling feature of contemporary industrial structures: the persistence of a skewed distribution of firm sizes, dominated by small firms, over all the industrial periods for which data can be estimated, approximately a Pareto distribution. This skewness exists not only for every industry and developed country, but has also persisted with remarkable stability over decades; no other phenomenon has in fact persisted in terms of this asymmetric distribution.

Recent developments thus indicate an increased skewness of the firm size distribution, that is, a quantitative trend superimposed on an otherwise remarkably stable characteristic of the evolution of national economies. Different overlapping economic forces or processes, which appear to have begun in the 1970s, may explain this trend. They can be summarised under three main categories (Keeble, 1990): recessionary pressures towards small business creation through unemployment and enforced entrepreneurship;

large firm fragmentation strategies and externalisation, which have induced growth in subcontracting by small manufacturing firms and provided new opportunities for small professional and business service enterprises (Keeble, Bryson and Wood, 1992; Bryson, Keeble and Wood, 1997); and increased differentiation of products and diversification and customisation of market demand, which have provided a host of new opportunities for small firms to exploit specialised market niches.

But even more important appears to be the qualitative change in the role of SMEs in economic growth which has accompanied the quantitative trend. What undoubtedly emerges from the period since the 1970s is the fact that SMEs can in many cases challenge large firms within the same competitive markets and lead the innovative process. This change in the relative position of SMEs became evident in the 1970s with the rise of Italy's small firm industrial districts in (mainly) traditional industries, which performed better than large firms within the same competitive markets because of the great flexibility allowed by the organisational and institutional arrangements developed by small local firms. These Italian industrial districts represent a kind of localised achievement of the division of labour where innovative capabilities and competitive efficiency have arisen from the organisational design established by the firms, arranged in networks of specialised subcontractors in vertically disintegrated processes of production (Brusco, 1982). Districts have developed as self-contained industrial systems of production composed of small firms working under the constraint of external markets for the final goods. They are characterised by high complementarity and synergy between firms and local institutions, which work together for local development. Local government, business associations, service centres, banks, collaborate to improve the co-ordination of local initiatives (Amin and Thrift, 1992). The nature of this internal co-ordination, far removed from hierarchical structures, has allowed Italian districts to overcome external challenges and perform better than large firms (Dei Ottati, 1998), because of the huge flexibility of the system which allows important capabilities of adaptation and innovation (incremental changes) in highly differentiated markets.

The evolution and persistence of these small firm industrial districts have challenged traditional theories of the firm and of industrial organisation, for example in relation to optimal firm size. But as Marshall (1890) classically, and Krugman (1991) more recently, have argued, the relevant unit of analysis of production and innovation may actually be geographic clusters of interdependent firms in which positive externalities emerge and knowledge spills over among firms.

A less immediately visible but equally important development has been the emergence of innovative high-technology SMEs and qualitative changes in advanced technology-based sectors. The 'self-contained' characteristics of Italian industrial districts with their striking clusters of small firms has naturally drawn attention to this phenomenon, whereas in technologically-advanced sectors similar patterns of small firm localisation have been much less visible, in Europe at least.[1] The development of such sectors has often arisen in an industrial context dominated by large firms and in a period of declining employment due to productivity restructuring. More emphasis has been put on such decline than on changes in the structure of the industries and the renewed role of SMEs in these sectors. These changes have also been difficult to trace because they have been even more challenging to traditional thinking than the rise of Italy's industrial districts. Indeed, that SMEs could emerge as engines of growth in sectors where technological change and innovation are the main competitive determinants might appear paradoxical. Innovation and technological change are supposed to require substantial R&D resources and, as empirical surveys have extensively shown, R&D is concentrated in large firms and proportional to size. In a dynamic context, a firm may admittedly engage in strategic behaviour to pre-empt the entry of rivals, by taking advantage of its current position in the market to engage in R&D so rapidly that it does not pay any rival to enter (Reinganum, 1989; Stiglitz, 1992). But clearly this requires that the existing firms are able to observe and react to potential entrants; the case of IBM which was unable, despite a huge R&D programme, to face competition from small entrants in a sector they had totally redesigned is significant. SMEs have been able to impulse changes without necessarily reproducing existing products or processes (Longhi and Raybaut, 1998).

Restructuring, flexible specialisation, division of innovative labour permitted by the changing nature of technological knowledge, and the emergence of science based activities, have all given numerous new opportunities to SMEs in technology-intensive rapid-growth sectors, as the shift in firm size structures documented above attests. But these general factors have not always been sufficient. The local contexts in which high-technology sectors are embedded have also played a basic role in the emergence of SMEs as engines of innovation and growth. Major universities for example have provided an important local source of knowledge and spillovers to innovative economic activities. Acs, Audretsch and Feldman (1994) have shown that spillovers from universities contribute more to the innovative activity of SMEs than to that of large firms, which are more concerned with industrial R&D spillovers. And Jaffe (1989) shows that such spillovers are very localised and fade over time and over space. Small events, like successful spin-offs from universities in a favourable cultural context, can trigger cumulative processes and growth based on these spillovers, as has occurred in Silicon Valley and Cambridge, UK, for example. Another facilitating element has been the presence of key 'enterprise generating' local firms, which have stimulated development through spin-offs, and spin-offs of spin-offs. The identification by local public actors of the importance of SMEs as vectors of change and sources of employment have also induced policies to promote their emergence and development in older industrial regions which are facing the necessity of adapting to new economic conditions. The result has been a steady growth in the share of SMEs in advanced economies, at least until the global recession of the early 1990s which will be discussed later (Figure 2.1). This process of growth in the importance of SMEs has gone hand-in-hand with other key interdependent evolutionary trends, mainly related to the location of economic activity and the organisation of the production and innovation processes.

2.2 Globalisation and Space

In recent years, the economic geography of advanced economies has undergone major changes. At the same point in time and for similar activities, high growth rates in some regions have coexisted with significant losses in other regions. Some regions have been able to absorb the repeated competitive and technological shocks which have characterised the last thirty years whereas others have declined. This spatial restructuring of industrial activity has been explained by existing inter-regional or national cost differences, which are seen as having triggered a continuing process of production restructuring and internationalisation of activities. This cost and market driven geography certainly exists and corresponds to the multinationalisation of production which developed in the 1970s, mainly as a result of the activities of large US multinationals. This phase was characterised by companies with strong headquarters and autonomous and self-contained regional operations without any necessary territorial roots, located in response to cost and market conditions. Local SMEs linked to such companies are particularly vulnerable since they can develop only as dependent subcontractors (Veltz, 1993).

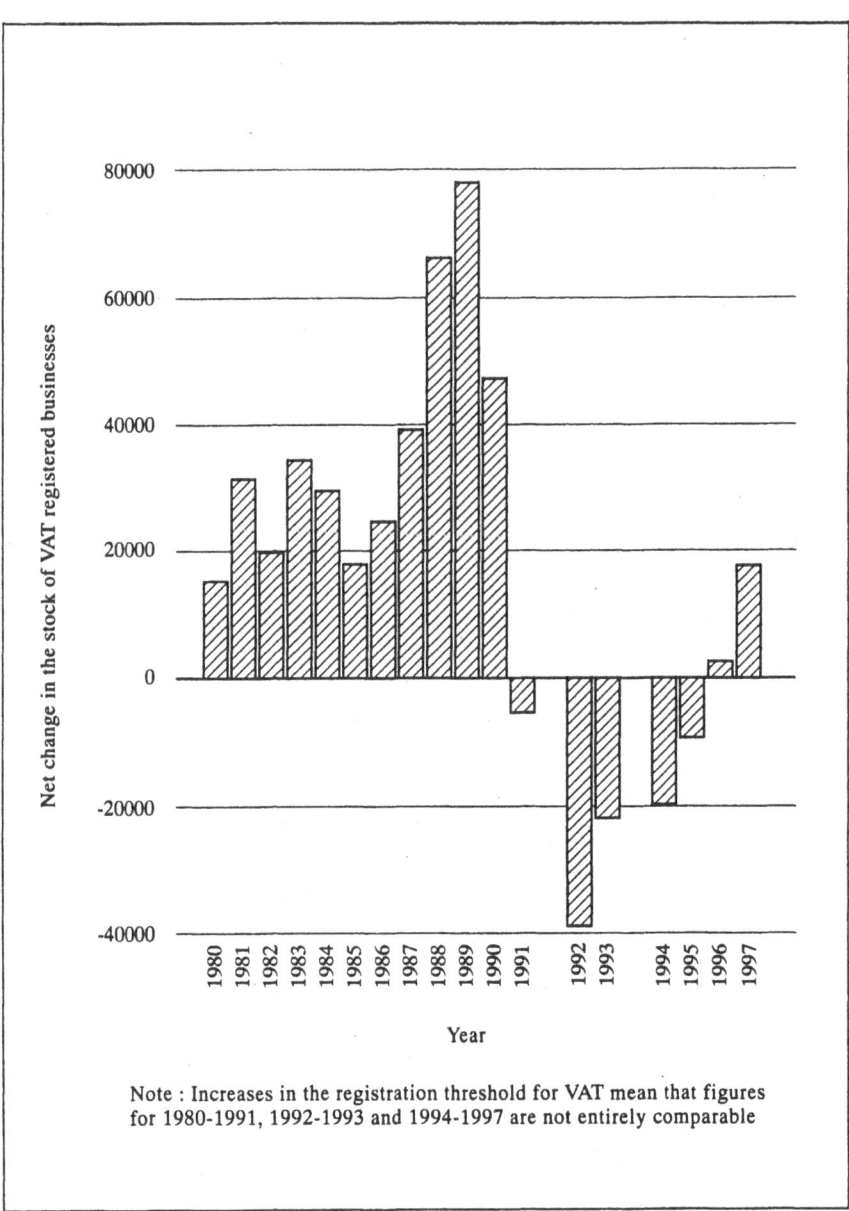

Figure 2.1   Growth, decline and resurgence in the UK's small firm population, 1980–97

Sources: Department of Trade and Industry, SME Statistics Unit.

The phase of globalisation which began in the 1980s corresponds to a totally different economic trend, whose focus is no longer on production costs and searching for new markets but on the innovative process itself. In the context of greatly increased competition and a resultant need for continuous innovation, globalisation for large firms no longer signifies unification of markets and fading of territories, but a global, interdependent approach to markets and global management of activities characterised by multiple territorial differentiation (Veltz, 1993). But paradoxically, the more the integration of activities has increased, the more the local has gained in importance. The emergence of new generic technologies has led to the emergence of new forms of production organisation and the development of innovative systems that go beyond the traditional frontiers of individual industries. In these 'cross-border systems of innovation', resource creation is the result not of the mere adding of technologies, but of the fusion of different technologies into a new technology (Imai and Baba, 1989). The process of innovation is no longer restricted within the boundaries of a single firm but brings together different technological capabilities and implies links between different actors, firms or industries, very often dispersed spatially (Gassman and Zedtwitz, 1999). The process of globalisation results in a globally integrated network, especially where new product conception and the innovation process are concerned. The more firm competitiveness depends on innovation-based production, and thus on different sets of competences and tacit knowledge, the more differences of location become important and meaningful (Gordon, 1996; Santangelo, 1999). Space is not neutral and the corollary of this process of globalisation is the increased importance of localisation.

2.3   Innovation and Localisation

The innovative process is not spaceless. On the contrary, innovation seems to be an intrinsically territorial, localised phenomenon, which is highly dependent on resources which are location specific, linked to specific places and impossible to reproduce elsewhere. These apparently reasonable propositions are in fact in total contradiction to the orthodox analysis of innovation which refers essentially to the adoption of a new but given technology (born from science) which is immediately efficient and available to every economic agent, and thus placeless.

A first step in reconsidering this traditional approach has been recent developments in the theory of localisation, which stress the existence of increasing returns in production (Arthur, 1990; Krugman, 1991); agglomeration economies and cumulative processes explain the concentration of activities in particular regions, as a consequence of 'small events', history, and path dependency of the economic process. These advances are an important step. But even more important than awareness of the consequences of the utilisation of increasing returns by firms, is understanding how these increasing returns are created. The problem is no longer one of locating activities, but of territorialisation. The analysis has then to organise itself according to the holy trinity defined by Storper (1997): Technology-Organisation-Territory. Technology is no longer given but is the result of a process of creation of new resources within firms, between firms, and between firms and other institutions. These institutional structures, which define the process of innovation, are deeply imbedded in particular territories. As stressed by Camagni (1995), space, traditionally considered as mere geographic distance, has to be replaced by territory, or relational space, defined through regional economic and social interactions. In the contemporary context of competition and innovation, the emerging 'leading regions' are those regions which have been able to develop endogenous processes of innovation, to implement processes of change and adaptation to the new environment. As noted earlier, even before large firms began to restructure their activities, particular territorially-embedded small firm systems, namely the Italian industrial districts, were able to find enough flexibility to adapt to the new conditions of competition and develop viable systems of innovation. This flexibility 'invented' by the territory has been adopted under different forms by the whole economic system.

The increasing uncertainty induced by increased competition and permanent innovation has led firms to adopt the organisational response of 'externalization-cum-disintegration' (Scott, 1988), and has reshaped the geographical structure of industry. Vertical interactions are increasingly most advantageously organised as external transactions between separate firms, while horizontal interactions with R&D or specialised high technology or service firms are increasing because of the increased complexity of the innovation process, which requires more face to face communication because of problems of tacit knowledge transfer and need for contracting. So firms have developed new interactive modes which are neither market nor hierarchy, but rather constitute network organisational forms characterised by 'relationship contracting' (Powell, 1990).

2.4 Network Regions and Innovative Milieux

This organisational change has involved major territorial upheavals and a process of economic development which is increasingly localised. The importance of R&D, and the need for specialised innovation services, have led firms to favour spaces with strong scientific and technological potential, where relevant information and research capabilities can be found. This dynamic has led to the emergence of particular metropolitan areas as strategic nodes of the contemporary economy. This trend is reinforced by the existence in these urban spaces of large labour markets with a wide range of specialised qualifications. Firms need access to increasingly specific human resources, as well as to external labour markets and competences derived from training in universities and engineering schools. This trend towards metropolitanisation offers firms a compromise between their need for stability, to secure the viability of the process of change, and their need for reversibility of choices, necessary in permanently innovation-driven economies (Veltz, 1993). The search for flexibility is thus a search for the reduction of risks associated with basic resources, mainly highly qualified workers and capabilities of mobility.

In these new arrangements, the 'model' of high technology territorial development which has emerged as the absolute reference point is Silicon Valley. The internal and external firm relationships prevailing in Silicon Valley enable it to be characterised as a 'network region' or metropolis (Gordon, 1996). This regional organisational form has evolved contemporaneously with globalisation of the innovation process. The structure of inter-firm relationships in Silicon Valley is distinguished by the prevalence of strategic cooperative alliances within but mainly outside the region, as the institutional base of technological innovation. Silicon Valley's internal dynamism is well-known, and, in addition to labour market and industry-research relationships, is fed by a continuous process of new firm creation, of spin-offs which are often established by workers from large firms (Storper, 1993). These spin-offs are created in an already institutionalised regional economy, not only technologically but also in terms of a specific professional culture, with strong specific competences and tacit knowledge. The emergence of new products and of specialised subcontracting enlarges the vertical and horizontal division of labour and sustains the growth of the local industrial complex and the internal market. Finally, the successful spin-offs usually sell themselves to large firms – to capture the innovative quasi-rents – and contribute to their technological renewal (Storper, 1993). These important processes of spin-off and sell-out contribute to the creation of regional collective learning, of specific locally-embedded competences, and are key characteristics of the region's 'innovative milieu', that is of 'the set of relations which unite the local production system, the set of actors and representation and the industrial culture, and which generates a localised dynamic process of collective learning' (Camagni, 1995: see also Camagni, 1991).

The industrial strategies born in Silicon Valley seem to have diffused world-wide. The process of globalisation of the innovation system has established metropolitan areas as preferential and interdependent places of creation of wealth. The non-metropolitan areas associated with this process of wealth creation are those local milieux or even micro-local milieux which have been able to join these global networks through their specific capabilities. Because it is built on innovation, the corollary of this process of globalisation-metropolitanisation led by large multinational firms appears to be :

- the emergence of local (or micro local) milieux as strategic elements of innovative networks. Regions are no longer substitutable as their involvement is based on scientific and technological knowledge;
- the role of high-technology SMEs as engines of development and sources of collective learning in metropolitan areas, and as key agents in the constitution and involvement of other areas in innovative networks.

2.5 The Regional Case Studies

The research carried out by the TSER Network has involved systematic analysis of different regional clusters of high-technology SMEs in order to highlight the processes of development characteristic of European technology-based regional economies in the 1990s. These case studies have investigated the role of high-technology SMEs in different spatial environments, from large metropolitan regions to localised high-technology production systems. Especially in Europe, globalisation does not automatically imply uniformisation, but involves a diversity of economic processes in different regions (Keeble, 1991; Longhi, 1998), a point of fundamental importance from a policy perspective. An important element of the research has thus been to understand how very different local systems of production have resulted in innovative milieux and collective learning. A wide range of different cases were therefore selected in different European Union countries, in order to cover different possible contexts: Cambridge, Oxford (UK), Milan, Pisa (Italy), Goteborg (Sweden), Munich (Germany), Grenoble, Sophia-Antipolis (France), Helsinki (Finland), Barcelona (Spain) and the Randstad (Netherlands).

The significance of our case studies for analysis of the regional concentration of high-technology activities in Europe is attested by the findings of the recent OST report (Barré, Laville and Zitt, 1998) on the 'Dynamics of S&T activities in EU regions'. The report is a major attempt to map European regions on the base of indicators of knowledge-based and technology-intensive activities. It classifies 445 EC regions (based on the NUTS2 and NUTS3 classification) through a cluster analysis run on three variables: productivity (GDP/population), technological density (number of European patents/population) and scientific density (number of scientific publications/population). Figure 2.2 maps those regions which are classified by the OST study in its top three (out of eight) clusters in terms of scientific and technological intensity.[2] All our regions are classified into the most technology-intensive cluster (type1), with the exception of Göteborg (type 2) and Barcelona (type 3). A similar finding emerges from even more recent research by Matthiessen and Schwarz (1999), which identifies eight of our ten case study regions as being amongst Europe's top 30 'scientific centres' measured by output of scientific, medical and engineering research publications 1994–96 relative to population, with Cambridge and Oxford coming first and second in this list.[3] This similarity in terms of exceptional science and technology intensity however conceals major differences in origins. The process of regional economic evolution, and even more the process of innovation, are historical and path dependent, and can be characterised by irreversibilities, increasing returns and lock-in. It is thus important to understand the history of these regions if we are also to understand their technological dynamics. We have thus built an a priori taxonomy, based on 'apparent' characteristics, accepting that the regions will also differ according to the nature of their collective learning processes and innovative milieu. Four classes are defined.

2.5.1 Industrial Regions : Munich, Grenoble and Göteborg

Europe's industrial regions have faced a deep crisis of restructuring, relocation of activities, and technological shocks since the 1970s, and many of them are still in transition towards new forms of activities. But the regions

32   *High Technology Clusters, Networking and Collective Learning in Europe*

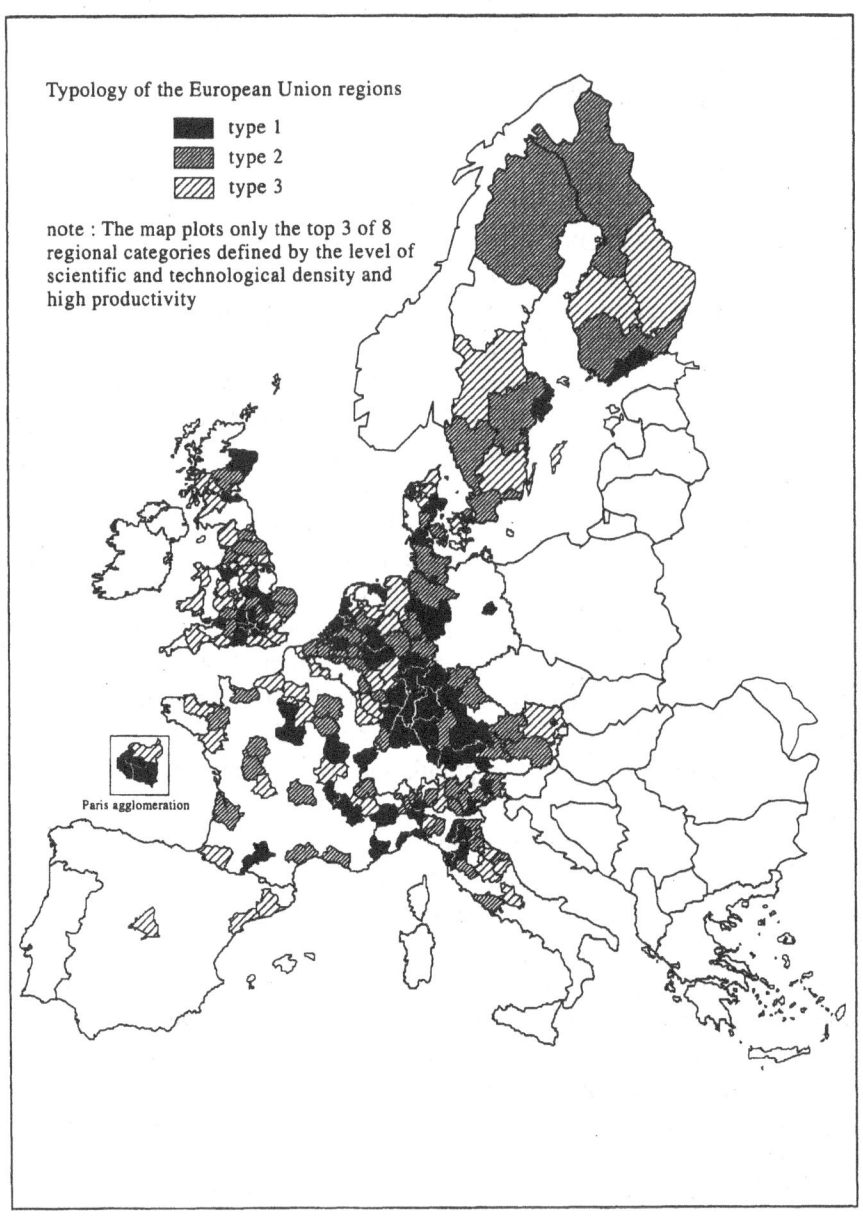

Figure 2.2   Europe's leading science and technology regions, 1995

Source: Barré et al. (1998), Map 1–4.

under consideration here have developed during their history collective capabilities allowing rapid adaptation to change.

Munich is an obvious case for inclusion as it ranks as the leading high-technology region in Germany whatever the indicator considered (Sternberg and Tamásy, 1998). The region can be characterised by the same institutional thickness and governance usually described for the more often analysed Baden-Wurttemberg (Cooke and Morgan, 1998), with a dense infrastructure of institutions dedicated to economic development (State or Land governments, chambers of commerce, etc) and numerous programmes dedicated to SMEs. These institutions support a dense network of inter-firm relationships, and of relations between universities and research institutes and firms. The region is characterised by the presence of large firms with headquarters in Munich which have often structured the region's industrial activity, in automobile manufacturing (BMW), aerospace (DASA), and electronics (Siemens). A newer sector, biotechnology, is being actively funded by the state government. Since the 1960s, Munich has also been the centre of Germany's strongly research-oriented armaments industry. The presence of highly innovative large firms with strong local linkages and of research intensive government programmes and funding have encouraged the development of a substantial population of innovative high-technology SMEs. Private sector industrial activity is complemented by major universities and research institutes. The Ludwig Maximilian University has developed important research programmes and technology transfer initiatives in medicine, information and communication technology, chemistry, energy and materials. It is involved in extensive interregional cooperation with SMEs, but spin off potential is still under-utilised.

Grenoble has an historic industrial tradition centred originally on hydroelectricity. The development of that industry at the beginning of the 20th century initiated the development of research and training activities. This long tradition has enabled Grenoble to overcome all the various shocks and crises it has subsequently faced. Its research and science capabilities have always been at the origin of new or renewed industrial activities, with new firm creation directly based on the findings of research institutes beginning in the 1950s. Firms like Merlin Gerin have been very important in the development of Grenoble's innovative milieu, and in building a local tradition of cooperation between research and industry. The AUG (University Alliance of Grenoble), which was created in 1947, still plays a major role in this respect. The emergence of new technologies was not contemporaneous with the decline of Grenoble's traditional industries, but began in the 1950s, and has resulted in the development of new materials, biotechnologies, electronics, and later software and computer services. The development of this potential is fundamentally linked to the region's public research and training potential, and the resultant local labour market of highly qualified employees. INPG (Institut National Polytechnique de Grenoble) brings together eight engineering schools, CNRS has established about ten research institutes, and Grenoble possesses three universities and CENG (Centre d'Etudes Nucléaire de Grenoble). This density explains the proliferation of SMEs since the 1970s. The development of the ZIRST-Meylan technopark clearly illustrates this dynamic process. But at the same time, only 70 of the region's 300 electronics SMEs are located in the ZIRST. High-technology development is really rooted in the whole local system.

The Göteborg region contains one of the strongest industrial concentrations in Sweden, with internationally-famous large firms but also numerous SMEs. This industrial tradition has resulted in the evolution of an innovative milieu characterised by different clusters of activity, which are distinguished by a high density of inter-firm relationships and links with local research institutions. Product, process and organisational innovations have endowed the milieu with considerable flexibility. Göteborg has different specialisations based on its large firms and their R&D laboratories, in automobile manufacturing, medical technology, telecommunications, information technologies, industrial electronics, and pharmaceuticals. Nevertheless, and notwithstanding the importance of large firms, Göteborg is also characterised by high rates of creation of new technology-based SMEs and the important role played by its universities in local economic development. Göteborg's Chalmers University of Technology has developed strong R&D collaborative links with industry and, together with its older companion Göteborg University, has generated approximately 350 direct technology spin-offs and some 115 indirect spin-offs. Chalmers has also established a science park and a range of innovation support organisations.

2.5.2 University-based Regions: Cambridge, Oxford and Pisa

Cambridge, Oxford and Pisa are local systems whose high-technology industrial dynamics have originated from the accumulated knowledge base of their universities. The Cambridge Phenomenon is undoubtedly the paradigmatic case, characterised by rapid growth since the 1970s of a diversity of technology-based SMEs. Though a region without a strong industrial tradition, Cambridge was exceptionally endowed with the resources necessary for the development of high-technology industries. For instance, it was an original source of development of computing in the UK, resulting in the establishment in the 1960s of the government-funded Computer Aided Design Centre. The industrial context has been built essentially through endogenous cumulative processes, with many direct or indirect spin-offs from the university. The university has de facto diagnosed the need for an industrial base to support some of its programmes and generate new synergies. Trinity College's 1970 decision to create the Cambridge Science Park was followed by St John's College in 1987 with its St John's Innovation Centre and Park (Reid and Garnsey, 1996). These initiatives have provided a strong image effect, but the Cambridge Phenomenon has been much wider than these initiatives, involving the whole urban region. Its evolution has been aided by the existence of social networks which have facilitated the creation of indigenous high-technology SMEs; until recently, the involvement of large firms has been limited, although these constitute major external markets for the innovation process.

In contrast to Cambridge, Oxford has an industrial history in which motor car and food manufacturing were leading sectors. This manufacturing tradition has influenced the development of high-technology industry, with two-thirds of local technology-intensive firms being manufacturing-based in the 1980s. A distinctive feature of the economy is the growth of large indigenous firms. The leading firm, Oxford Instruments, has had a major impact both on the development of the region and the growth of numerous spin-off firms by former employees. The university's initial impact on local high-technology growth was much less than in Cambridge. Indeed, Oxford University has traditionally seen its research mission as national rather than local; though developing a range of programmes oriented to high-technology sectors, relevant partners were generally large national firms or multinationals. In Oxford, local technology transfer has been seen until recently as the responsibility of non-university organisations such as the Oxford Trust, explicitly created to promote local science and technology-based firms. Nonetheless, university spin-offs and linkages are an important component of the evolution of this regional cluster, as is the role of the local scientific and professional labour market created, in part, by the university.

Pisa is an important university town in Italy, with three scientific universities (State University, Scuola Normale (physics), Scuola Superiore S. Anna) and many units of the National Research Council. An important knowledge base has been built around computer sciences, electronic engineering, pharmaceuticals, robotics, and aerospace. Industrial development has taken the form of research centres of large international firms (Olivetti, Hewlett-Packard), and of technology-based SME spin-offs from external large firms, which have chosen to locate in Pisa because of its existing knowledge base. Some spin-offs from the universities exist, but without a strong cumulative development process. These spin-offs are more explained by pull factors (personal success and scientific prestige, concretisation of ideas) than by push factors (dissatisfaction with academic life or impossibility of developing the projects in the universities) (Camagni, 1995). In this context, Pisa is a very open university-based region, with most of the final users of Pisa high-technology services being located outside the region.

2.5.3 Technopoles and Peripheral Regions: Sophia-Antipolis and Barcelona

The Sophia-Antipolis project is representative of both technopolis-based and peripheral region development. Indeed, the project was created in the 1970s in a vacant space, a region – the French Riviera – without any industrial or university tradition. The project was aimed at promoting a new type of development based on a 'sunbelt' effect, and was launched after the unforeseen location in the region of IBM's and Texas Instruments' research centres. The project has benefited from the involvement of France Télécom and the availability of an important and advanced telecommunication infrastructure, and the location of public research centres from the CNRS (Centre National de la Recherche Scientifique), INRIA (Institut National de Recherche en Informatique et Automatique), and Ecole Supérieure des Mines. Two important trends have generated a significant industrial base for the project: the national decentralisation of activities from Paris, and the multinationalisation of firms, mainly US, which have used Sophia-Antipolis as a European hub for their activities (with specialisation in computer sciences, telecommunications, and pharmaceuticals). In contrast to the previous cases, economic development has been purely exogenous, and emerged without any involvement from the nearby university of Nice or the location or creation of high technology SMEs. Only since the mid-1980s has the university and a significant number of SMEs begun to participate in local economic development, which has therefore taken on a more endogenous form. The creation of local business associations, involving large and small firms and training and research institutes, has accelerated this process.

Barcelona has passed through a long economic crisis, accentuated by the decline of the Mediterranean ports, but in recent years has experienced a significant renewal of its industrial dynamism. Still, the level of R&D expenditure and the development of its high-technology sectors are weak compared to the other regions in our sample. Its industrial dynamism has been supported by strong public policies of attracting foreign investment; FDI is historically important in Catalonia, which has some characteristics of a peripheral region. Since the mid-1980s, Japanese investments have been dominant, often through acquisition of local firms. The main sectors are automobile, components, and pharmaceutical manufacturing, in which foreign firms have been attracted by a highly qualified workforce. Recently, a new policy to develop links between research centres and industry has been implemented through the University Reform Law, aimed at ending the traditional separation between universities and companies and allowing universities to conduct contract research. Institutions have been created to assist this process and foster contacts between the different elements of the Science-Technology-Industry system. Barcelona has also established a science park, but few spin-offs yet exist and technology transfer to industry is often limited to consultancy.

2.5.4 Metropolitan Regions: Milan, Helsinki and the Randstad

Europe's leading metropolitan regions are the locus of global networks of innovation, and Milan, Helsinki and the Dutch Randstad are involved in this process. Milan is one of the birthplaces of Italian industry, and with Genoa and Turin constitutes the 'Italian Triangle'. It is noteworthy however that with the restructuring crisis and the emergence of new models of innovation, Genoa, dominated by public sector firms, and Turin, with its marked industrial specialisation, have been dragged down in a long crisis of transition whereas Milan has found enough flexibility to adapt and enough capabilities to develop new S&T activities and attract innovative multinationals. Out of the three cities, only Milan has all the characteristics of a global metropolis: numerous skills in diversified high qualifications, specialised professional and business services, financial services, and training and institutes of research (Politecnico). Since the 1960s, the growth of new high-technology sectors in the north-east of Milan has resulted in the 'Milan innovative field', comprising a dense interlinked complex of large and small firms specialising in computers, electronic equipment, telecommunications, pharmaceuticals, and R&D services.

Helsinki has also faced a major crisis of reconversion after the collapse of USSR and its formerly important industrial market. Recovery has been achieved through a shift from traditional to high-technology industries and the development of a new model of innovation. Like Milan, the region's assets have been important: a highly-qualified labour market and wide range of skills, financial services, a strong concentration of scientific and technological universities and research institutes, headquarters of the large Finnish firms, and a strong industrial culture and identity. These assets have been reinforced by an active public policy, at both state and city levels, promoting high-technology development. Uusimaa county, of which Helsinki is the core, is particularly active in developing programmes to foster management skills of entrepreneurs in technology-based enterprises. Different science parks and technology centres have been created. The Otaniemi science park, founded in 1984, claims to be the largest centre of technology-based activities in northern Europe. Its Innopoli technology centre houses 200 firms (160 high-technology SMEs) engaged in information technology and telecommunications. The Helsinki science park specialises in biotechnology. Each park is endowed with university departments and technological institutes. There are thus strong synergies between public institutions, S&T bodies and large and small firms.

The Randstad region, in the west of the Netherlands, contains the four largest cities in the country: Amsterdam, The Hague, Rotterdam, and Utrecht. The region is the most densely populated in Europe. But high densities also characterise the incidence of firms, universities and research institutes. Wherever firms locate in Netherlands, they are close to a university or institute. The region is well endowed with knowledge-based services (S&T, financial, professional) and large, often multinational, firms. One of the main characteristics of the Randstad is its concentration of high-technology SMEs. However, only a relatively small proportion of these are manufacturing firms; some 80 per cent are engaged in providing high-technology services, mostly involving information technologies, especially software (medical and food technologies are secondary specialisations). The Randstad's labour markets supply a wide range of highly-qualified workers and spin-off potential is high. Vertical or horizontal links between SMEs or with research centres do exist, but many strategic links related to the innovation process are external to the region, reflecting well established national or international networks.

2.6 The Evolution of High-Technology Regional Clusters in the 1990s

As argued earlier in this chapter, the main contemporary European regional clusters of high technology SMEs emerged in the 1970s and 1980s as the result of rapid technological change, rising rates of small firm creation and entrepreneurship generally, and organisational changes adopted by large firms and local economic systems faced with recession, intensifying competition, and the need for continuous innovation. Despite some similarities (in regional residential attractiveness and quality of life for scientists and entrepreneurs, and communication and telecommunication infrastructures), a key characteristic of the process was the great diversity of forms and mechanisms of emergence of the clusters: endogenous versus exogenous processes, a large firm-driven process of development versus dynamics of SME creation and entrepreneurship, public investment in science versus infrastructure. The high path-dependency of the regional evolutionary process has inevitably resulted in some continuing differences in regional adaptation to the economic stresses and opportunities of the 1990s. Nonetheless, most of the clusters have now arguably developed the key characteristics of an innovative milieu, while there is also evidence of some convergence in evolutionary processes and patterns during the past decade.

After the rapid economic and small firm growth of the 1980s, the early-1990s were characterised by a severe European-wide recession. In the UK, for example, the total small firm population, proxied by the stock of VAT-registered businesses, declined by nearly 100,000 or 5.6 per cent between 1991 and 1995 because of a steep rise in failure rates and decline in start-ups (see Figure 2.1). Similar trends were recorded by a majority of EU countries (European Commission, 1996, p.58). This acute recession inevitably impacted to some extent on Europe's high technology SME clusters. Sophia-Antipolis, for example, recorded a significant decline in the rate of employment growth (though employment still grew) between 1991 and 1994 (Figure 2.3), while the Cambridge region appears to have experienced a fall in its population of high-technology SMEs after 1993, notwithstanding continuing growth in employment. Europe's technology-intensive SMEs almost certainly however weathered the early-1990s recessionary storm appreciably better than other small firms. Reid and Garnsey (1996, 1997), for example, have conclusively established that Cambridge high-technology SMEs have exhibited significantly lower failure rates than their low-technology counterparts.

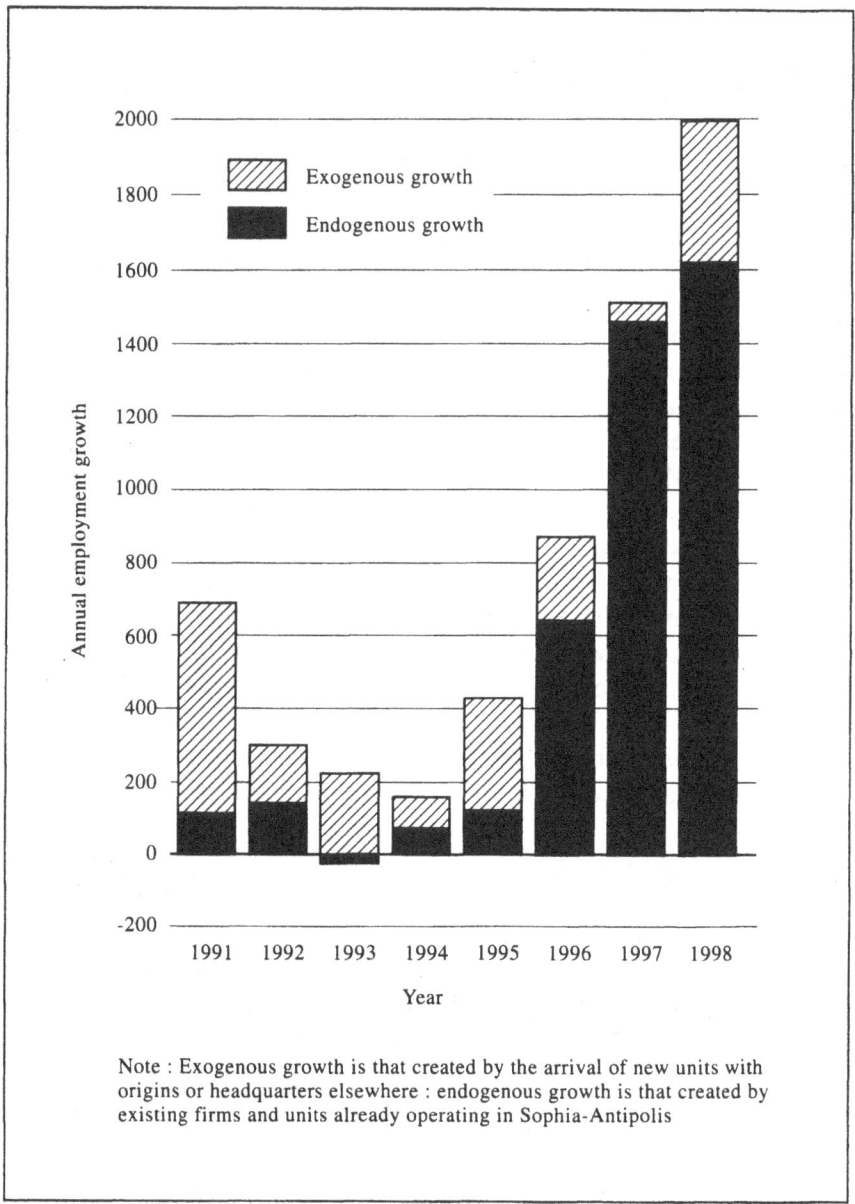

Figure 2.3 Exogenous and endogenous employment growth in Sophia-Antipolis, 1991–97

Source: SAEM Sophia-Antipolis

With European recovery from recession since the mid-1990s, most of the high-technology SME clusters studied have experienced a marked resurgence of growth. Sophia-Antipolis' net annual employment growth soared from under 200 to 1990 jobs 1994–98 (Figure 2.3), industrial firms in Helsinki's high technology-focussed Uusimaa county grew by 6 per cent (stock) and 13 per cent (turnover) 1993–94, Göteborg's new technology-based firm (NTBF) formation rate increased each year 1994–97, and high-technology employment in the Cambridge region grew by 32 per cent 1991–98. Over the whole 1989–96 period, the Munich region generated the largest volume of new technology-intensive SMEs of all German regions. Many of Europe's technology-based regional SME clusters thus appear to have been growing significantly in the late 1990s, in a dynamic and possibly self-sustaining evolutionary process.

Given the great historical diversity of regional cluster origins and evolutionary paths described above, it is noteworthy that this 1990s resurgence of technology-intensive SME growth arguably shows some signs of regional evolutionary convergence, in terms of sectoral structures, growth processes involving spin-offs and large firms, the role of universities and public research laboratories, and the development of new forms of regional collective enterprise.

Sectorally, 1990s trends in a number of the study regions appear to involve increased regional diversification of sectoral structures, rather than increasing regional specialisation upon only one or a few sectors. Regions are thus converging in the sense of becoming somewhat more rather than less similar in sectoral composition. Thus the pattern of NTBF creation in the Göteborg technology-based regional cluster indicates that this is evolving away from its traditional specialisation upon advanced mechanical engineering products and towards softronics (software and electronics), mechatronics (combining mechanical and electronic technologies in new ways), and technology-intensive services (Lindholm Dahlstrand, 1997). In Sophia-Antipolis, small firm resurgence is focussed on a diversity of new sectors such as telecommunications, multi-media, information technology and computer applications (Longhi, 1999), a trend which is also evident in the Cambridge region.

In addition, sectoral convergence is also occurring through a rapid process of blurring of traditional divisions between sectors, with a coalescing and combining of different technologies (computers, telecommunications, biotechnology, medical and health products, for example). Sectorally diversified clusters appear to offer particularly favourable opportunities for such coalescence and combination, often in the form of new spin-offs. Increased diversification however also involves in some cases the rise of new 'micro-clusters', as in the growth in the Cambridge region of new biotechnology, telecommunications, and internet, multimedia and computer applications firms. It also appears to be associated with a marked increase in importance of high-technology services, relative to high-technology manufacturing. Four out of every five extra high-technology jobs created in the Cambridge cluster in the 1990s have been in technology-based services (computer, software and information technology applications, R&D consultancy, telecommunications), not manufacturing. In the Göteborg case, it is in technology services that there has been 'the fastest growth of technology-based activity in the region in the 1990s' (Lindholm Dahlstrand, 1997).

Convergence in growth processes relates to the balance between spin-offs and new firms created by local entrepreneurs (endogenous growth),[4] and external large firm investment through takeovers and new research laboratories (exogenous growth). In the 1990s, some clusters historically dominated by small local spin-offs, such as Cambridge, have been experiencing high rates of indigenous firm acquisition (Garnsey and Cannon-Brookes, 1993) and new inward investment by large multinationals (Microsoft), including investment in 'embedded laboratories'. The latter are private sector firm R&D units located actually on university or hospital sites, sometimes even sharing buildings with university laboratories, as with Microsoft's new west Cambridge facility. Multinational companies are thus now showing considerable interest, often for the first time, in participating in the growth of Europe's high-technology SME clusters, by investment, takeover and local collaborative networking. In contrast, the formerly large firm-dominated 'satellite platform' (Markusen, 1996) of Sophia-Antipolis has witnessed the striking new phenomenon of the growth of small spin-offs from both public research institutes and large firms: as Figure 2.3 shows, by 1998 over 80 per cent of net employment growth was endogenous, from spin-offs and existing local firms, not from new inward investment (see also Longhi, 1999). New technology-based firm creation rates have also risen in the Göteborg region, another traditionally large firm-dominated industrial economy.

A further recent trend arguably involving some degree of convergence concerns the role of regional knowledge centres, in the form of universities and public research institutes. In the university-based high-technology clusters studied by the Network (Oxford, Cambridge, Pisa), growth and maturing of the cluster has by the 1990s inevitably diminished the originally dominant influence of the key universities involved. In the Cambridge case, for example, a recent survey of local high-technology SMEs rated research links with Cambridge university only eleventh, in terms of importance for the firm's development, out of a list of 19 possible development influences (see Keeble et al. 1999). In a number of the non-university based clusters, however, the role of local universities (Chalmers University of Technology in Göteborg, Helsinki University of Technology, University of Nice/Sophia-Antipolis) appears to have intensified and assumed greater importance in the 1990s. This reflects the development of deliberately pro-active technology transfer, new firm spin-off, and research collaboration policies, and increased importance of universities in providing highly-qualified scientific and research staff for the regional high-technology labour market. Universities in the university-based clusters studied have also become increasingly pro-active with regard to stimulating and promoting the growth of the local high-technology community, as for example by encouraging the establishment of multinational 'embedded laboratories', as described above.

Finally, there is independent evidence in several regions of the simultaneous emergence during the 1990s of new forms of 'regional collective enterprise', in terms of new collective partnership initiatives designed to promote and improve regional marketing, publicity, networking, business support and infrastructure provision. These are frequently business-led but also involve other key regional actors in the form of local government, public business support and training agencies, universities and public research institutes. They appear to reflect a growing realisation of the potential benefits of local networking, including networking between large and small firms, of the need for collective endeavour to overcome growing infrastructure and other local constraints on the continuing growth of successful clusters, and of the importance of a cluster's global image and regional marketing in a rapidly-shrinking internet-driven world. Examples of new regional collective enterprise include the recent establishment in Cambridge of such local networks and organisations as Cambridge Connect and the Greater Cambridge Partnership (see chapter 8), and in Sophia-Antipolis of the Telecom Valley Association and EuroSud 155 (Longhi, 1999).

2.7   Why New 1990s Trends?

At least four different explanations can be put forward to account for these recent changes in the dynamics and evolutionary trajectories of European high-technology regional clusters. These focus on the impact of radical technological change, new or enhanced characteristics of production organisation by high-technology SMEs, new characteristics of globalisation of large multinational firms, and cluster maturation and growth to achieve 'critical mass' and self-sustaining dynamism.

First, the 1990s has witnessed radical and global technological change, with exceptionally rapid and research-driven technological developments in the biotechnology and information and telecommunication technologies sectors, amongst others. These sectors' high research-intensity, their extraordinary speed of technological change, and explosive growth of demand for their products and services (internet applications, computer games, etc) helps explain the proliferation of small, new entrepreneurial firms in these sectors in a number of the study regions. The growth of the European biotechnology industry, based on remarkable technological developments in gene technology, cloning, bioscience and drug delivery systems, exemplifies this trend, with the number of European 'entrepreneurial bioscience' companies identified by Ernst and Young (1995, 1999) in its annual survey more than doubling in the last four years, from 485 in 1995 to 1,080 in 1995.[5] The 'vast majority of these companies employ fewer than 50 people' (Ernst and Young, 1997), while they are strikingly clustered geographically in particular European regions such as Cambridge, Stockholm, Helsinki, Paris, Berlin and Munich, as Figure 2.4 shows.

Rapid technological development and associated explosive growth in demand also help explain exceptional growth in numbers of computer, internet and information technology service SMEs, in this case coupled with low barriers to entry and the evolution of numerous specialised market niches which small firms can target successfully. As Cane (1999) points out, 'the explosive growth of the internet has opened up myriad opportunities for start-ups to develop innovative solutions to problems that did not exist a decade ago'. In the United Kingdom, for example, one in eight of all new enterprises registered in the whole economy in 1997 were in computer services, while the number of businesses with at least one employee in computer software, services and applications (SIC 72) soared by no less than 128 per cent, or 48,182, in only four years end-1993 to start-1997 (DTI, 1995, 1998, 1999). No less than 95.7 per cent of these are small firms employing less than 10 employees. In contrast, in more traditional high-technology manufacturing sectors where economies of scale are essential,

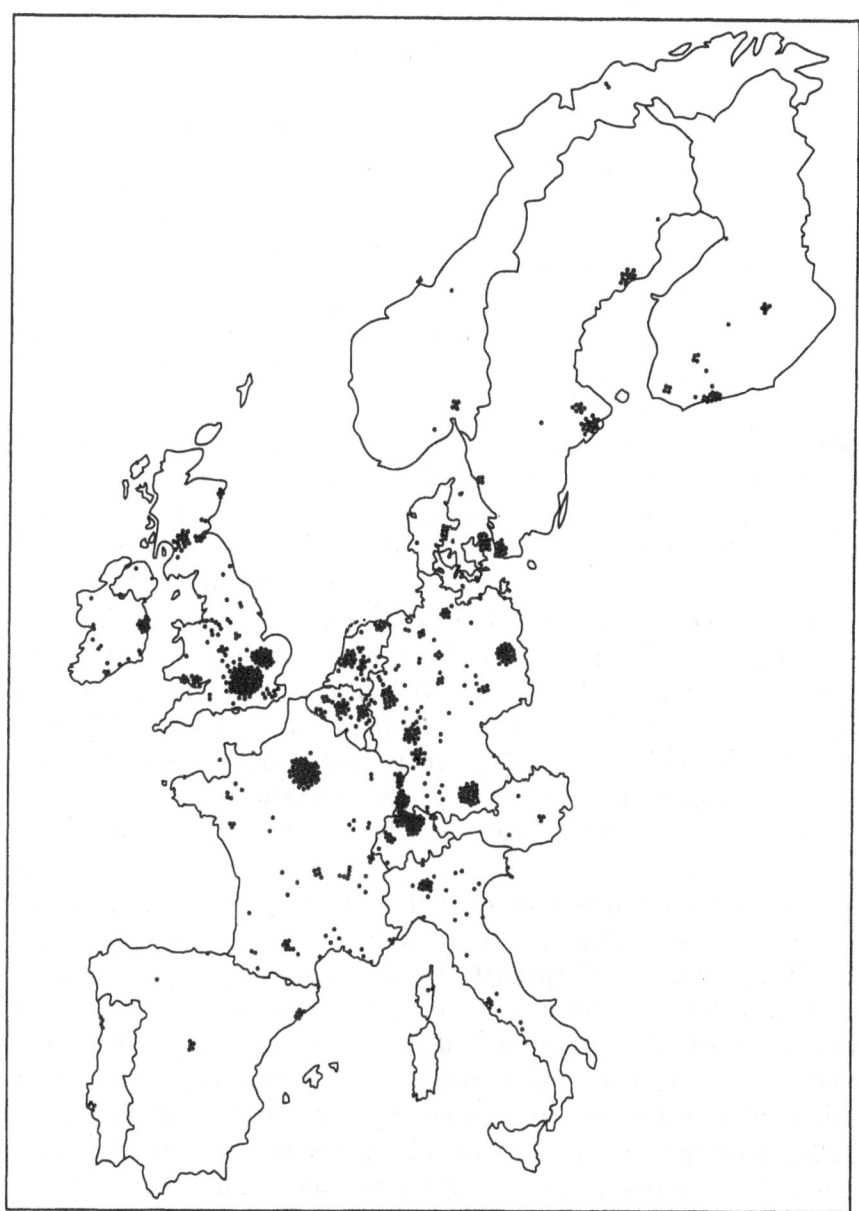

Figure 2.4 Regional clustering and the European entrepreneurial life sciences sector, 1999

Source: Ernst and Young, 1999, p. 7.

such as consumer electronics, semi-conductors and aerospace, large firms remain dominant, a pattern also observable in some new rapidly-growing high-technology manufacturing industries, such as mobile telephones. Small R&D-intensive firms do however often play a very important role in these sectors – and the regions, such as Munich, in which their large firms operate – as specialised subcontractors and service providers (Sternberg and Tamásy, 1999).

A second very important explanation for these regional evolutionary trends lies in new or enhanced characteristics of the organisation of production by high-technology SMEs. Recent Cambridge University research (Hughes, 1998) shows that by the late 1990s, R&D-intensive SMEs in the UK differ significantly from more conventional small firms in exhibiting much higher rates of technological innovation, intensities of networking and collaboration with other firms and organisations, niche market orientation, levels of globalisation, and use of external – and often local – information and business advice. These differences are documented in Table 2.1, reproduced from data in Hughes (1998), which are in turn derived from the Cambridge University ESRC Centre for Business Research's 1997 national survey of 2,500 manufacturing and professional and business service SMEs (Cosh and Hughes, 1998). High-technology firms were defined in this research on a firm-by-firm, not sectoral, basis in terms of the firm's above-average R&D intensity (R&D expenditure equal to or greater than 4.0 per cent of sales in the past financial year).

Hughes' research demonstrates conclusively that by the late 1990s, R&D-intensive SMEs in the UK exhibit significantly different patterns of organisation of production compared with 'conventional' manufacturing and professional service SMEs in all the respects listed above (Table 2.1). First, and within a context in which employment and turnover size does not differ significantly between the groups of high-technology and conventional SMEs surveyed, the former are far more innovative than their 'low-technology' counterparts, with nearly twice as high a proportion of both manufacturing and service firms reporting that they had developed and introduced product innovations during the previous three years, and were planning further product innovation in the next three years. The differences with regard to radical or novel innovations (products or services which are entirely new to the industry in which the firm operates) are even more striking, with seven out of every ten manufacturing high-technology firms, and two-thirds of service high-technology firms, reporting developing radical innovations. While not perhaps surprising, these empirical findings

Table 2.1 New or enhanced characteristics of organisation of production by high-technology SMEs in Great Britain

	Manufacturing SMEs		Professional and business service SMEs	
	High-technology	Conventional	High-technology	Conventional
Innovation				
• % product innovators	92.3	52.5	87.5	45.9
• % novel product innovators	71.2	27.5	65.2	25.0
• % planning future product innovation	96.8	59.5	94.9	53.2
Networking and collaboration				
• % with collaborative or partnership arrangements	52.6	23.2	71.7	34.9
• % collaborating with customers	70.7	54.0	50.7	38.0
• % collaborating with universities/HEIs	30.5	13.2	16.9	9.4
Niche market orientation				
• mean number of serious competitors	6.4	16.4	8.4	19.3
Globalisation				
• mean number of serious overseas competitors	2.8	1.3	2.1	1.5
• % of serious competitors overseas	37.5	14.0	33.7	6.9
• importance of export growth as business objective	48.4	21.3	29.0	8.9
• importance of overseas market share as business objective	41.8	16.0	26.6	8.8
Number of firms	175	1010	104	745

Table 2.1 cont'd

Accessing External Advice

High-technology SMEs in both manufacturing and services are significantly more likely than conventional SMEs to seek external advice on:

- business strategy
- management organisation
- marketing, market research and public relations
- product and service design
- new technology
- staff recruitment, training and human resource development

High-technology SMEs in both manufacturing and services are significantly more likely than conventional SMEs to seek external advice from:

- accountants and lawyers
- customers
- management and business consultants
- business friends and relatives
- local government business advisory and training organisations

Notes:
1) Networking and collaboration refers to firms 'entering into formal or informal collaborative or partnership arrangements with other organisations in the last 3 years'
2) Importance of business objectives measured as % of firms rating this as a 'very significant' or 'crucial' objective during the past 3 years
3) Product innovation refers to introduction to the market of a 'new or significantly improved product' during the last 3 years: novel product innovation refers to an innovation which is 'new to your firm *and* to your industry'
4) External advice refers to 'external sources from which you have obtained business advice (as distinct from basic information) in the last 3 years'

Source: Hughes, 1998: see also Cosh and Hughes, 1998.

highlight the importance of high-technology SMEs for successful European, national and regional innovation, and innovation policies.

A second very important difference is that to develop their innovative products, UK high-technology SMEs are more than twice as likely as conventional small firms to engage in collaborative networking via formal and informal partnerships with other firms and universities. Such collaborative networking appears to have been growing in importance amongst European small firms in recent decades, but is peculiarly characteristic of innovative high-technology SMEs. Nearly three out of every four high-technology service firms, and over half of the manufacturing high-technology firms, report being involved in collaborative networks. Most of these are with customers, suppliers or subcontractors, but a significant minority of manufacturing high-technology SMEs (Table 2.1) collaborate with universities and research institutes. Other Cambridge work suggests that many of these networks and partnerships are also likely to be local or regional, as in the Cambridge region (Keeble et al., 1999). This distinctive current characteristic is clearly very important in understanding the growth of the European regional clusters of high-technology SMEs investigated in this book.

Thirdly, while most small firms exist to serve specialised or niche markets (Kitson and Wilkinson, 1996), the number and range of which have grown significantly over the last three decades (Small Business Research Centre, 1992, pp. 18–19) both manufacturing and service high-technology SMEs are even more targeted at such niche markets than are conventional firms. The former report fewer than half the number of serious competitors, an indicator of segmented and specialised markets used in a number of recent studies (Pratten, 1991, ch. 6), than conventional firms. The reason for this is often linked with technological leadership and innovativeness, and the development of advanced technologically-based products or services for which initially at least only a limited and specialised set of customers exists.

Fourthly, the very small number of such customers in any one national market may well enforce globalisation via exporting or licensing from a very early stage in a technology-based firm's evolution (Keeble et al., 1998). Equally, globalisation and success in global markets offers possibilities of much greater growth and profits, as in the world's most successful high-technology regional cluster, Silicon Valley, where 'over three quarters (of innovative high-technology SMEs) are oriented by design to international markets' (Gordon, 1992). Table 2.1 thus also reveals that European high-technology SMEs are significantly more oriented to global customers and confronted by global competitors than their conventional counterparts. In a globalising world economy, many small technology-based firms are actively pursuing policies of targetting foreign customers and markets, particularly in other European countries and North America. Again, this global orientation renders them potentially even more important as basic income-generating activities for national and regional economies than their conventional counterparts, many of which exist only to serve local or regional 'derived' demand.

The final characteristic of the organisation of production by European high-technology SMEs in the 1990s which differentiates them from other small firms, and which is important in understanding the role of regional clustering in their recent evolution, is that they make significantly greater use of external business advice of many different kinds – on business strategy, management organisation, marketing, market research and public relations, product and service design, the implementation of new technology, and staff recruitment and training (Table 2.1). This undoubtedly partly reflects the fact that such firms are characteristically established and managed by highly-educated and professionally- and managerially-qualified researchers and entrepreneurs (Whittaker, 1999, pp. 5–6), who are inherently outward-looking and aware of the need for specialised external advice and know-how. Equally, most of this advice involves high-trust relationships (Bennett and Robson, 1999) and is provided and delivered locally, by accountants, lawyers, consultants, business associates and local business support and training agencies. These findings fully corroborate those on particular European high-technology clusters by members of the TSER Network. Over 80 per cent of Cambridge region high-techology SMEs, for example, use local accountants, banks, legal services, personnel and recruitment agencies, and design and printing services for 50 per cent or more of their external business service needs (Keeble et al., 1999). SMEs in the regional clusters studied are exceptionally embedded in both local and global collaborative, customer and advisory networks, both types of network being apparently important for competitive success.

New or enhanced characteristics of production organisation by Europe's high-technology SMEs are parallelled by the third probable explanation of recent regional cluster evolutionary trends, namely new characteristics of the globalisation process of large firms. As outlined earlier, the 1990s appear to be witnessing a major re-orientation and redefinition of large multinational firm strategies with respect to European regional clusters of technology-based SMEs. Globalisation and frenetic technological change appear to be driving many large firms, and especially firms in technology-intensive sectors, to place technological innovation at the heart of their competitive strategies. In turn, successful innovation demands access to locational specificities in terms of specialised regional research and professional labour markets, university and research institute technology competencies, and existing networks of innovative high-technology SMEs. As Santangelo (1999, p. 16) has recently demonstrated for ICT (information and communications technology) industries in the European context, 'the success of [multinational firm] innovative activity appears to be increasingly more embedded in local centres of expertise', since 'the location of R&D laboratories in a dynamic regional environment enables the whole MNC to source abroad knowledge complementary to its technological path'. The result is the recent and continuing attraction to many of the European clusters studied of multinational R&D laboratories, as with the recent establishment in Cambridge of Microsoft's first-ever R&D laboratory outside North America, and in Sophia-Antipolis of research units of Siemens, British Telecom and Toyota. Microsoft have even coupled this decision with the establishment of a dedicated Cambridge venture capital fund to promote new local technology-based start-ups. Embedded large firm laboratories are thus increasingly seeking to tap into local knowledge networks and SME communities, thus further encouraging the growth of regional SME clusters. Regional differences in SME and university R&D and technology competences are thus becoming more important in the large firm globalisation process than regional differences in costs or infrastructure. This new trend is likely further to sustain and stimulate the growth of Europe's high-technology SME regional clusters into the twenty-first century.

The final and perhaps simplest explanation of the new 1990s trends is the growth over time in size of many European SME regional clusters to achieve some degree of 'critical mass' and hence self-sustaining dynamism via local spin-offs and spillovers. While none can emulate the huge scale of US examples such as Silicon Valley, the Network's research does suggest that by the late 1990s, most of the clusters studied had grown to a sufficient size to generate regional internal economies of scale in terms of specialised labour markets, skills and services, and to achieve significant levels of continuing spin-offs and new firm creation. The result in most cases has been the development of a significant capacity for 'regional collective learning' and knowledge development (see chapters 7 and 8), focussed on SMEs and their links with universities, public research institutes and large firms. New regional partnerships and collective enterprise initiatives are another example of the impact of maturation and growth in critical mass. At the dawn of the twenty-first century, processes of networking, innovation, knowledge development and collective learning within European regional clusters of technology-intensive SMEs appear to lie at the heart of these regions' undoubted economic success.

Notes

1. This is in contrast to the USA, whose high-technology SME clusters such as Silicon Valley (California), Orange County (California), Minneapolis (Minnesota) and Austin (Texas) have attracted world-wide interest and attention.
2. We are most grateful to Professor Barré and his colleagues for permission to reproduce part of their original map as our Figure 2.2.
3. The regions involved, in order of ranking, are Cambridge (1), Oxford-Reading (2), Helsinki (8), the Randstad (10), Munich (11), Göteborg (13), Milan (18) and Barcelona (29).
4. Recent convergence between the Cambridge and Oxford high-technology clusters in this respect is also noted by Garnsey and Lawton Smith (1998, p. 444).
5. The 1999 report gives a total of 1,178, but this appears to include 98 companies in Israel, which are therefore excluded from the total here. Note also that the map in the 1999 Ernst and Young report contains errors in the plotting of firms in Sweden, which have been rectified in Figure 2.4. We are most grateful to Mr David Hales of Ernst and Young International for permission to reproduce their original data in our map.

References

Acs, Z.J and Audretsch, D.B. (1993), *Innovation and Small Firms*, MIT Press, Cambridge, Mass.

Acs, Z.J., Audretsch, D.B. and Feldman, M.P. (1994), 'R&D Spillovers and Recipient Firm Size', *Review of Economics and Statistics*, vol. 76, pp. 336–40.

Amin, A. and Thrift, N. (1992), 'Neo-Marshall Nodes in Global Networks', *International Journal of Urban and Regional Research*, vol. 16, pp. 571–87.

Arthur W.B. (1990), 'Silicon Valley's Locational Cluster : When Do Increasing Returns Imply Monopoly ?', *Mathematical Locational Sciences*, vol. 19, pp. 235–51.

Audretsch, D.B. (1995), *Innovation and Industry Evolution*, MIT Press, Cambridge, Mass.

Barré, R., Laville, F. and Zitt, M. (1998), *The Dynamics of S&T Activities in the EU Regions*, Observatoire des Sciences et des Techniques (OST), Paris.

Bennett, R.J. and Robson, P.J.A. (1999), 'The Use of External Business Advice by SMEs in Britain', *Entrepreneurship and Regional Development*, vol. 11, pp. 155–80.

Birch, D.L. (1981), 'Who Creates Jobs ?', *The Public Interest*, 65, Fall, pp. 3–14.

Brusco, S. (1982), 'The Emilian Model: Productive Decentralisation and Social Integration', *Cambridge Journal of Economics*, vol. 6, pp. 167–84.

Bryson, J., Keeble, D. and Wood, P. (1997), 'The Creation and Growth of Small Business Service Firms in Post-Industrial Britain', *Small Business Economics*, vol. 9, pp. 345-360.
Camagni, R. (1991), 'Local 'Milieu', Uncertainty and Innovation Networks: Towards a New Dynamic Theory of Economic Space', in R. Camagni (ed.), *Innovation Networks: Spatial Perspectives*, Belhaven Press, London, pp. 121–43.
Camagni, R. (1995), 'High-Technology Milieux in Italy and New Reflections about the Concept of "'Milieu Innovateur'", paper presented at European Workshop on High-Technology Enterprise and Innovative Regional Milieux, Cambridge, 3–4 March 1995.
Cane, A. (1999), 'High-tech Specialist Happy to be Target Purchase', *The Financial Times*, 18 August 1999, p. 1.
Capello, R. (1998), 'Collective Learning and the Spatial Transfer of Knowledge: Innovation Processes in Italian High-Tech Milieux', in D. Keeble and C. Lawson (eds), *Collective Learning Processes and Knowledge Development in the Evolution of Regional Clusters of High-Technology SMEs in Europe*, ESRC Centre for Business Research, University of Cambridge, Cambridge, pp. 19–37.
Cooke, P. and Morgan, K. (1998), *The Associational Economy: Firms, Regions, and Innovation*, Oxford University Press, Oxford.
Cosh, A. and Hughes, A. (eds) (1998), *The State of British Enterprise*, ESRC Centre for Business Research, University of Cambridge, Cambridge.
Dei Ottati, G. (1998), 'The Remarkable Resilience of the Industrial Districts of Tuscany', in H-J. Braczyk, P. Cooke and M. Heidenreich (eds), *Regional Innovation Systems*, UCL Press, London.
DTI (Department of Trade and Industry) (1995), *Small and Medium Sized Enterprise (SME) Statistics for the United Kingdom, 1993*, Small Firms Statistics Unit, DTI, Sheffield.
DTI (Department of Trade and Industry) (1998), *Business Start-ups and Closures: VAT Registrations and De-registrations in 1997*, Statistical Press Release, Small Firms Service, DTI, Sheffield.
DTI (Department of Trade and Industry) (1999), *Small and Medium Enterprise (SME) Statistics for the United Kingdom, 1998*, SME Statistics Unit, DTI, Sheffield.
Dosi, G. and Salvatore, R. (1992), 'The Structure of Industrial Production and the Boundaries between Organizations and Markets', in M. Storper and A.J. Scott (eds), *Pathways to Industrialization and Regional Development*, Routledge, London, pp. 171–92.
Ernst and Young (1995), *European Biotech 95: Gathering Momentum*, Ernst and Young International, London.
Ernst and Young (1997), *European Biotech 97: A New Economy*, Ernst and Young International, Stuttgart.
Ernst and Young (1999), *Ernst and Young's European Life Sciences 99: Sixth Annual Report, Communicating Value*, Ernst and Young International, London.
European Commission (1996), *Enterprises in Europe: Fourth Report*, Office for Official Publications of the European Communities, Luxembourg.
European Network for SME Research (1994), *The European Observatory for SMEs, Second Annual Report*, EIM Small Business Research and Consultancy, Zoetermeer.
Garnsey, E. and Cannon-Brookes, A. (1993), 'The "Cambridge Phenomenon" Revisited: Aggregate Change among Cambridge High-Technology Companies since 1985', *Entrepreneurship and Regional Development*, vol. 5, pp. 179–207.
Garnsey, E. and Lawton Smith, H. (1998), 'Proximity and Complexity in the Emergence of High-Technology Industry: The Oxbridge Comparison', *Geoforum*, vol. 29, pp. 433–50.

Gassman, O. and von Zedtwitz, M. (1999), 'New Concepts and Trends in International R&D Organization', *Research Policy*, vol. 28, pp. 2-3.
Gonzales-Benito, J., Reid, S. and Garnsey, E. (1997), 'The Cambridge Phenomenon Comes of Age', *Research Papers in Management Studies*, The Judge Institute of Management Studies, University of Cambridge, WP 22/97.
Gordon, R. (1992), 'PME, Réseaux d'Innovation et Milieu Technopolitain : la Silicon Valley', in P. Maillat and J.-C. Perrin (eds), *Entreprises Innovatrices et Développement Territorial*, GREMI-EDES, Neuchâtel, pp. 195-220.
Gordon R. (1996), 'Industrial Districts and the Globalization of Innovation: Regions and Networks in the New Economic Space', in X. Vence-Deza and J.S. Metcalfe (eds), *Wealth from Diversity*, Kluwer, Dordrecht, pp. 103-34.
Hughes, A. (1998), 'High-Tech Firms and High-Tech Industries: Finance, Innovation and Human Capital', paper presented at Conference on SMEs and Innovation Policy: Networks, Collaboration and Institutional Design, ESRC Centre for Business Research, University of Cambridge, Cambridge, 13 November 1998.
Ijiri, Y. and Simon, H.A. (1977), *Skew Distribution and the Size of Business Firms*, North Holland, Amsterdam.
Imai, K. and Baba, Y. (1989), 'Systemic Innovation and Cross-Border Networks', paper presented at International Seminar on Science, Technology and Economic Growth, OECD, Paris.
Jaffe, A. (1989), 'Real Effects of Academic Research', *American Economic Review*, vol. 79, pp. 957-70.
Keeble, D. (1988), 'High-Technology Industry and Local Environments in the United Kingdom', in P. Aydalot and D. Keeble (eds), *High-Technology Industry and Innovative Environments: The European Experience*, Routledge, London, pp. 65-98.
Keeble, D. (1990), 'Small Firms, New Firms and Uneven Regional Development in the United Kingdom', *Area*, vol. 22, pp. 342-245.
Keeble, D. (1991), 'Core-Periphery Disparities and Regional Restructuring in the European Community of the 1990s', in H.H. Blotevogel (ed.), *Europäische Regionen im Wandel*, Dortmunder Vertrieb fur Bau- und Planungsliteratur, Dortmund, pp. 49-68.
Keeble, D., Bryson, J. and Wood, P. (1992), 'Entrepreneurship and Flexibility in Business Services: The Rise of Small Management Consultancy and Market Research Firms in the UK', in C. Caley, E. Chell, F. Chittenden and C. Mason (eds), *Small Enterprise Development*, Paul Chapman, London, pp. 43-58.
Keeble, D. and Bryson, J. (1996), 'Small Firm Creation and Growth, Regional Development and the North-South Divide in Britain', *Environment and Planning A*, vol. 28, pp. 909-34.
Keeble, D. and Lawson, C. (eds) (1998), *Regional Reports*, ESRC Centre for Business Research, University of Cambridge, Cambridge.
Keeble, D., Lawson, C., Lawton Smith, H., Moore, B. and Wilkinson, F. (1998), 'Internationalisation Processes, Networking and Local Embeddedness in Technology-Intensive Small Firms', *Small Business Economics*, vol. 11, pp. 327-42.
Keeble, D., Lawson, C., Moore, B. and Wilkinson, F. (1999), 'Collective Learning Processes, Networking and 'Institutional Thickness' in the Cambridge Region', *Regional Studies*, vol. 33, pp. 319-32.
Keeble, D. and Wever, E. (1996), 'Introduction', in D. Keeble and E. Wever (eds), *New Firms and Regional Development in Europe*, Croom Helm, London, pp. 1-34.

Kitson, M. and Wilkinson, F. (1996), 'Markets and Competition', in A. Cosh and A. Hughes (eds), *The Changing State of British Enterprise*, ESRC Centre for Business Research, University of Cambridge, Cambridge, pp. 22–31.

Krugman, P. (1991), *Geography and Trade*, MIT Press, Cambridge, Mass.

Lindholm Dahlstrand, A. (1997), 'Technological Specialisation in the Göteborg Region', in D. Keeble and C. Lawson (eds), *Networks, Links and Large Firm Impacts on the Evolution of Regional Clusters of High-Technology SMEs in Europe*, ESRC Centre for Business Research, University of Cambridge, Cambridge, pp. 67–74.

Longhi, C. (1998), 'La Dynamique des Espaces Urbains : Innovation et Marché du Travail', *Les Annales de la Recherche Urbaine*, pp. 134–45.

Longhi, C. (1999), 'Networks, Collective Learning and Technology Development in Innovative High Technology Regions: The Case of Sophia-Antipolis', *Regional Studies*, vol. 33, pp. 333-342.

Longhi, C. and Raybaut, A. (1998), 'Free Competition', in R. Arena and C. Longhi (eds), *Markets and Organization*, Springer-Verlag, Heidelberg, pp. 95–124.

Markusen, A. (1996), 'Sticky Places in Slippery Space: a Typology of Industrial Districts', *Economic Geography*, vol. 72, pp. 293–313.

Marshall, A. (1890), *Principles of Political Economy*, Macmillan, London.

Matthiessen, C.W. and Schwarz, A.W. (1999), 'Scientific Centres in Europe: An Analysis of Research Strength and Patterns of Specialisation Based on Bibliometric Indicators', *Urban Studies*, vol. 36, pp. 453–77.

Powell, W.W. (1990), 'Neither Market nor Hierarchy: Network Forms of Organization', *Research in Organizational Behaviour*, vol. 12, pp. 295–336.

Pratten, C. (1991), *The Competitiveness of Small Firms*, Cambridge University Press, Cambridge.

Reid, S. and Garnsey, E. (1996), 'Incubator Centres and Success in High-Technology Firms: The Work of the St. John's Innovation Centre', Cambridge University Engineering Department and Judge Institute of Management Studies, ESRC Innovation Programme report.

Reid, S. and Garnsey, E. (1997), 'High-Tech – High Risk? High Tech Firms Are Not Risky and Need More Funding', *New Economy*, vol. 4, pp. 131–5.

Reinganum, J.F. (1898), 'The Timing of Innovation: Research, Development and Diffusion', in R. Schmalensee and R. Willig (eds), *Handbook of Industrial Organization*, North-Holland, Amsterdam, pp. 849–908.

Santangelo, G.D. (1999), 'Inter-European Regional Dispersion of Corporate Research Activity in Information and Communication Technology: The Case of German, Italian and UK Regions', *University of Reading, Department of Economics, Discussion Papers in International Investment and Management*, Series B, No. 268.

SBRC (Small Business Research Centre) (1992), *The State of British Enterprise*, Small Business Research Centre, University of Cambridge, Cambridge.

Scott, A. (1988), *New Industrial Spaces: Flexible Production Organisation and Regional Development in North America and Western Europe*, Pion, London.

Simon, H.A. and Bonini, C.P. (1958), 'The Size Distribution of Firms', *American Economic Review*, vol. 42, pp. 607–17.

Sternberg, R. and Tamásy, C. (1998), 'Munich', in D. Keeble and C. Lawson (eds), *Regional Reports*, ESRC Centre for Business Research, University of Cambridge, Cambridge.

Sternberg, R. and Tamásy, C. (1999), 'Munich as Germany's No. 1 High Technology Region: Empirical Evidence, Theoretical Explanations and the Role of Small Firm/Large Firm Relationships', *Regional Studies*, vol. 33, pp. 367–77.

Stiglitz, J.E. (1992), *The Meaning of Competition in Economic Analysis*, Stanford University, mimeo, 41 pp.

Storper, M. (1993), 'Regional "Worlds" of Production: Learning and Innovation in the Technology Districts of France, Italy and the USA', *Regional Studies*, vol. 27, pp. 433-455.

Storper, M. (1997), *The Regional World*, Guilford Press, New York.

Veltz, P. (1993), 'D'une Géographie des Coûts et une Géographie de l'Organisation: Quelques Théses sur l'Evolution des Rapports Entreprises/Territoires', *Revue Economique*, vol. 4, pp. 671–84.

Whittaker, H. (1999), 'Entrepreneurs as Co-operative Capitalists: High Tech CEOs in the UK', *ESRC Centre for Business Research, University of Cambridge, Working Paper* 125.

3 Regional Institutional and Policy Frameworks for High-Technology SMEs in Europe

CHRISTINE TAMÁSY AND ROLF STERNBERG

3.1 Introduction

Innovation processes are influenced by many factors and can develop in various institutional and policy frameworks, which together can be regarded as 'systems of innovation'. Systems of innovation are seen as supra-national, national or sectoral (Nelson, 1993; Breschi and Malerba, 1997). Here the emphasis will be on regional systems which many researchers regard as useful models for a better understanding of the production and distribution of knowledge in local economies, and for policy formulation (see, for example, Cooke, 1998). The definition of 'regional innovation systems' and their scope varies somewhat in the literature (see chapter 1, section 1.6). However, certain broad features of regional innovation systems have been highlighted. First, particular institutional factors can distinguish one region from another.[1] These may be related to the structure and quality of a region's industrial research, its R&D institutions (universities, public laboratories, etc.), its infrastructure for supporting technology transfer, and its financial institutions, which all may help the innovation process. A second group of factors relate to a region's public policy, both its form – either direct (R&D funding) or indirect (taxation) – and its targeting (fields of R&D, encouragement of high-technology SMEs). In this regard, innovation appears to be less the product of the individual firm and more the result of networked institutional and policy frameworks (Kline and Rosenberg, 1986). Proximity between industries and academic research centres raises the probability of interaction and collective learning processes so that knowledge spill-over and the transfer of tacit knowledge cannot be explained without taking regional contexts into account.

This chapter analyses the institutional and policy frameworks for high-technology development in 10 European regions: Barcelona (Spain), Cambridge and Oxford (England), Göteborg (Sweden), Grenoble and Sophia-Antipolis/Nice (France), Helsinki (Finland), Milan (Italy), Munich (Germany) and Utrecht/Randstad (Netherlands).[2] The second section will discuss general aspects of the technology-oriented environment and various contexts. The third section will provide an overview of national and regional trends of research and development in Europe. Section four will examine technology transfer and knowledge infrastructure in the regions surveyed. Section five will analyse the role of technology policy. Section six will conclude.

3.2 General Aspects of the Technology-Oriented Environment and Various Contexts

One of the most striking characteristics the systems of innovation approaches have in common is their emphasis on the role of 'institutions' (Edquist, 1997, p. 25). This can be divided into eight parts:

- higher education institutions (e.g., universities, advanced technical colleges);
- state-owned or partially state-owned research institutions outside the university;
- science parks or innovation centres;
- other institutions dedicated to technology transfer, SMEs, innovation;
- key entrepreneurs or key persons;
- large firms with (more or less) strong activities in industrial research and development;
- technology fairs;
- technology policy.

Figure 3.1 shows specific ways in which regional institutional frameworks can influence the performance of high-technology SMEs. This model is somewhat formal; nevertheless, it indicates the flows of information, know-how, advice etc. that can significantly affect the technology-oriented environment. Firm characteristics act as filters, since businesses, even those within the same regional innovation system, tend to have specific needs and developmental bottlenecks dependent upon their

Regional Institutional and Policy Frameworks 59

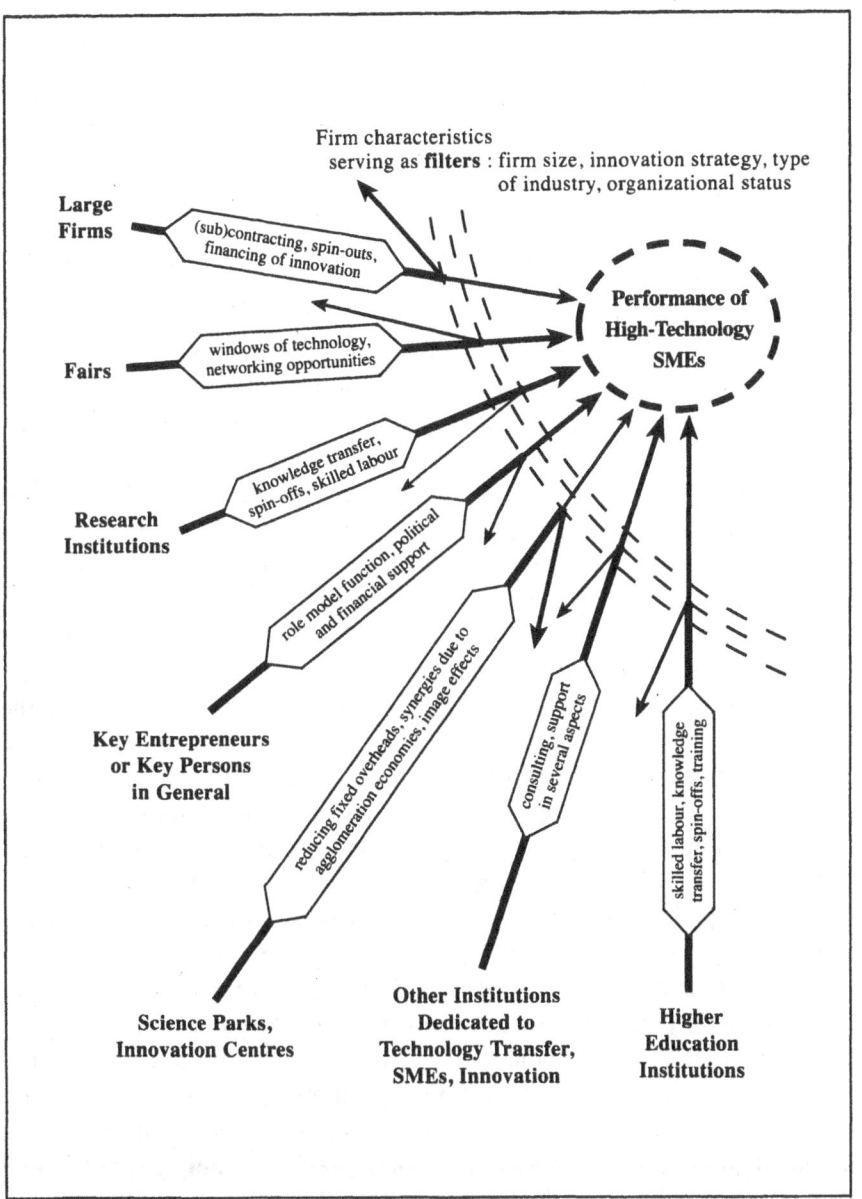

Figure 3.1 Impacts of regional institutional frameworks on high-technology SMEs' performance

size, innovative strategy, branch of business, etc. Firm-specific characteristics are important considerations when assessing the impact of technology policy on the performance of high-technology SMEs. For example, whether or not an SME is a branch plant of an enterprise with its headquarters elsewhere will influence its propensity to co-operate with local research institutions or innovation support organisations.

For the purpose of this study, technology policy will be defined as: 'involving government intervention in the economy with the intention of affecting the process of technological innovation' (Stoneman, 1987, p. 36). Such policies can be expected to have a variety of direct and indirect effects. For example, government policy on R&D expenditures (e.g., commissioned research) and non-market demands for technology-intensive products (e.g., with respect to defence policy) influence the performance of high-technology SMEs or the availability and quality of technology transfer and knowledge infrastructure. At the same time the purchasing areas of the various participants may be geographically diverse (Figure 3.2). National policy generally dominates. However, this is being increasingly complemented by institutions on the local, regional and supra-national levels. Technology policy in Germany provides a representative example of this. Currently, innovation centres are the most popular instruments for policymakers at the local level. Regionally, the Federal State of Bavaria supports the introduction of new technologies in SMEs. Nationally, the Federal Ministry of Research and Technology as well as the Federal Ministry of Defence sponsor R&D projects of civil and military-strategic importance in both large and small firms. At the supra-national level the European Union encourages the participation of SMEs in EU-funded R&D programmes (Technology Stimulation Measure for SMEs). Similar divisions in spatial jurisdictions exist in other network member countries.

In the regions favouring high-technology SMEs, goals of technology policy do not deliberately include spatial effects. In region A (Figure 3.2), for example, only the regional government explicitly pursues spatial goals. The regional effects of the other two policy levels may be considerable but they are usually unintended. Therefore, the following statement applies not only to the majority of EU technology programmes, but also to most individual government programmes: 'they are "ostensibly aspatial" and mainly promote already existing potentials; and in doing so they implicitly promote agglomerations' (Amin and Pywell, 1989, p. 472). In short, the unintended spatial impacts of technology policies are far greater than the intended ones (see Sternberg, 1996).

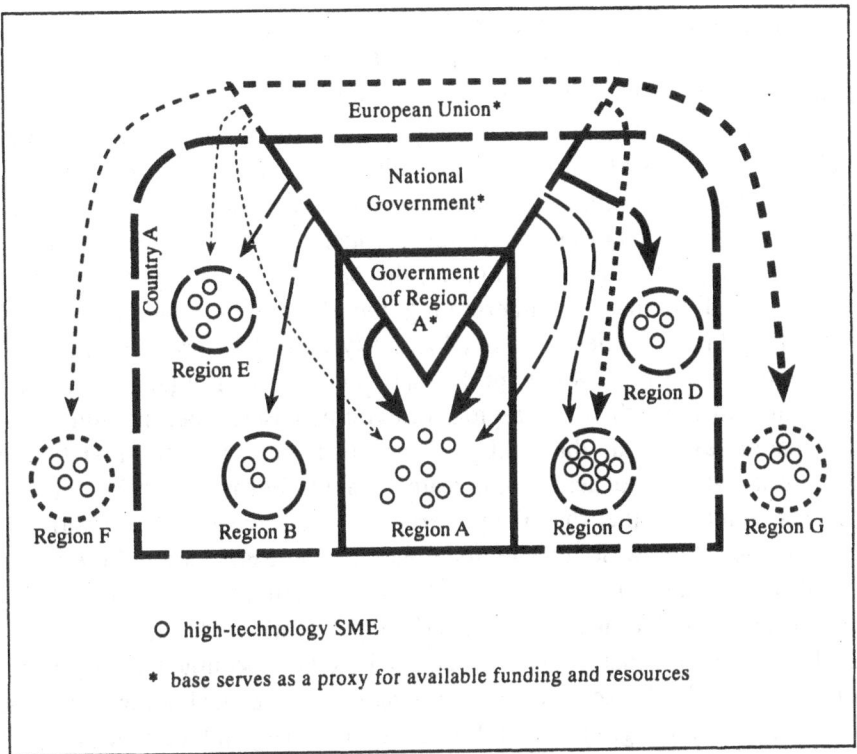

Figure 3.2 Impacts of technology policies on high-technology SMEs' performance

3.3 National and Regional Trends of Research and Development in Europe

In this section, institutional differences in Europe are reflected in regional and national comparisons. The hypothesis is that the institutional and policy frameworks in network regions studied are shaped to a great extent by regional and national specialities in research and development. The intention is to emphasise the regional and national differences in technological development, whereby the state (government, higher education) and private businesses are the key players in innovation systems. For comparison two of the most frequently used indicators are used – namely the absolute size and relative significance (as a per cent of Gross Domestic Product (GDP)) of R&D expenditures.

Research and development comprise creative efforts undertaken on a systematic basis to increase the stock of knowledge. R&D expenditures include all funds used to advance research and development within a reported unit or sector of the economy. This includes current expenditures such as employment costs or expenditures on materials, plus capital expenditures on buildings, equipment etc. R&D expenditure data therefore provides us with a set of indicators very frequently used to analyse scientific and technological progress. However, the data is not wholly comparable, because of differences in interpreting the definitions, disparate survey methods and peculiarities of national R&D systems. Dividing up the area of study into regions is especially problematic in the case of organisations with a presence in more than one region. In extreme cases, when no additional information exists, all the R&D expenditures must be attributed to the regions in which the headquarters are located (Eurostat, 1997). Finally, R&D is an input, rather than the preferred output, indicator of innovation.

Statistics show wide inter-regional variation in R&D expenditure in the European Union. For instance, in 1993, Sweden allocated the highest percentage of GDP to R&D activities (3.20 per cent), while Greece brought up the rear with a mere 0.48 per cent of its GDP accounted for by R&D (Eurostat, 1997). These figures give a national level technology gap[3] of around 1:7. At the regional level, Baden-Württemberg had the highest R&D to GDP ratio (3.89 per cent), while Dytiki Macedonia had the lowest at 0.04 per cent. Figure 3.3 shows that a small number of technology-intensive regions dominate the European R&D system. From the available data,[4] nine regions stand out with a level of expenditure greater than 2.5 per cent of GDP, namely Baden-Württemberg, Berlin, Ile de France, East Anglia, Midi-Pyrénées, Bavaria, Vienna, the South East and Bremen (NUTS 1 regions except Midi-Pyrénées and Vienna). In short, the technological gap has been much greater at the regional than at the national level.

Figure 3.3 also shows the differences in R&D expenditures of the three main sectors (business enterprise, higher education and government) in the network member countries. In all eight countries except Spain, the business enterprise sector accounted for more than the other two sectors combined in 1995 (Netherlands, 1994). By contrast, in Spain the higher education sector and government sector combined accounted for more than half of R&D expenditures. The highest level of R&D expenditure in the government sector occurred in France, the only country in the European Union where it is greater than the higher education sector. However, it is important to add that R&D expenditure within the business enterprise sector is highly

Regional Institutional and Policy Frameworks 63

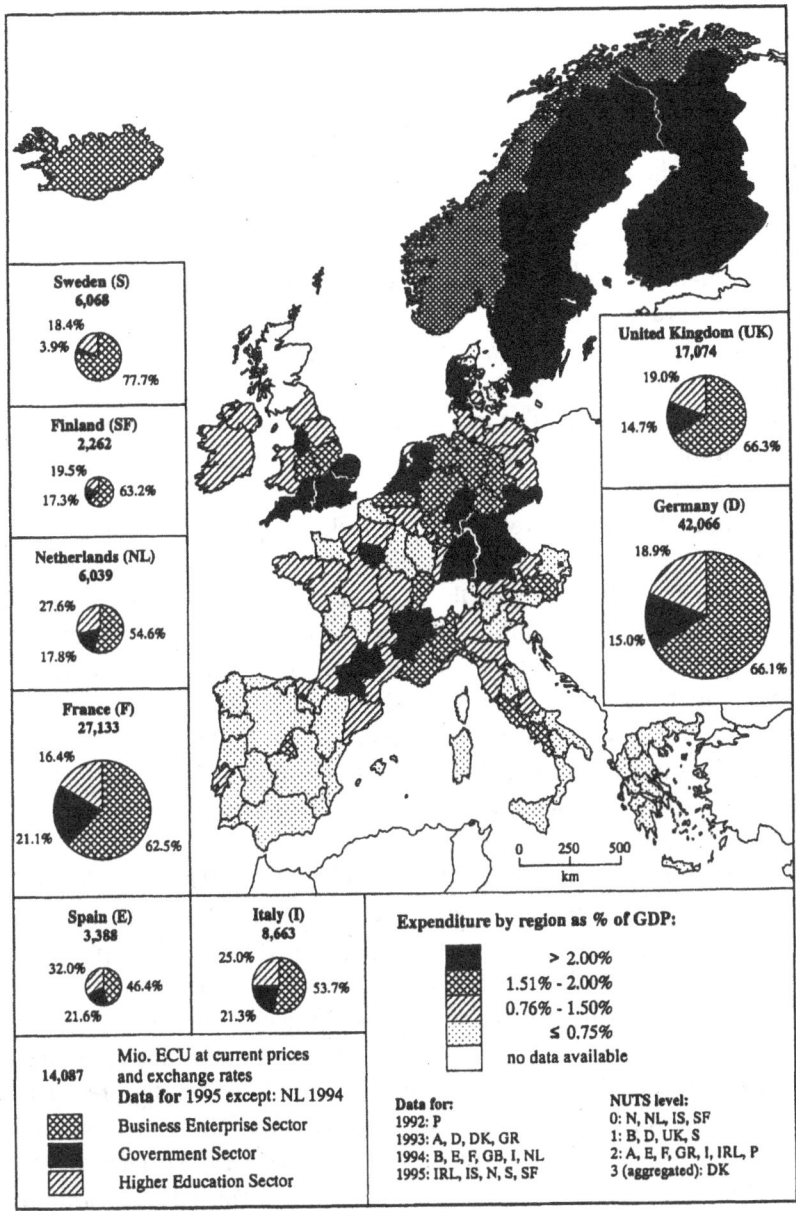

Figure 3.3 EU R&D expenditure by region

Source: Eurostat, 1997.

concentrated in manufacturing groups. For example, in Sweden, manufacturing accounted in 1995 for 84 per cent of the business sector's R&D expenditures and within manufacturing seven large firms – Ericsson, Volvo, Saab, Astra, Scania, Sandvic and Incentive – accounted for as much as 78 per cent of the manufacturing total (NUTEK, 1997).

3.4 Assessment of Technology Transfer and Knowledge Infrastructure

Technology transfer and knowledge infrastructure are key elements of regional innovation systems. The main feature of most regions studied is the high concentration of R&D institutions as well as innovation support organisations.[5] However, the capacity to influence the development of linkages is limited in a few regions by a lack of resources (Oxford, for example, has one of the smallest innovation support teams in the UK; see Lawton Smith, 1998). Regional disparities in the technological infrastructure influence technology-based SME growth to an extent depending upon the technical orientation and academic level of R&D institutions, which can provide personnel for R&D, qualified graduates, knowledge and technology transfer and a creative atmosphere.

In addition, successful research institutions act as incubators for new firms. For example, Chalmers University of Technology generated 240 direct university spin-off companies between 1969 and 1993, most of them in the Göteborg region (Lindholm Dahlstrand, 1997). The founding of technology and innovation centres are based on expectations of spin-offs and (in most regions studied) are very popular examples of property initiatives designed to facilitate technology transfer between universities and industry. These institutions have attracted much attention and publicity, but they cater for only a minority of local firms. The most important advantage of innovation centres is their real estate function, while the regional economic effects (creation of new jobs, intensification of technology transfer) are normally small (see, for example, Tamásy, 1999). The science park of Sophia-Antipolis is, however, a special case. It 'started from nothing and has progressively structured in this region a real industrial capacity that represents around 15,000 jobs today' (Longhi and Quéré, 1997, p. 219).

Non-market institutions such as universities and local government organisations provide continuity and store knowledge relevant for R&D industries. These are relatively stable elements in regional systems of

Table 3.1 Assessment of institutional frameworks for high-technology SMEs in 10 European regions

Regions	Institutions						
	Higher Education Institutions	Research Institutions	Science Parks, Innovation Centres	Technology Transfer Institutions	Key Entrepreneurs or Persons	Large Firms	Fairs
Barcelona	◐	◐	◐	○	○	◐	◐
Cambridge	●	◐	●	○	◐	○	○
Göteborg	◐	◐	◐	◐	○	◐	◐
Grenoble	◐	●	●	◐	◐	◐	○
Helsinki	●	●	●	◐	○	●	○
Milan	◐	●	◐	○	●	○	●
Munich	◐	◐	○	○	●	●	◐
Oxford	●	◐	◐	◐	●	◐	○
Utrecht/Randstad	◐	◐	◐	○	○	◐	◐
Sophia-Antipolis	◐	●	●	○	●	◐	○

The assessment is based upon the relative significance of the institutions in each region (but not between regions)

● strong impact on high-technology SMEs
◐ medium impact on high-technology SMEs
○ almost no impact on high-technology SMEs

innovation (Garnsey, 1998). Nevertheless, academic institutions and industry have in essence, very different value systems, objectives and organisational structures. Interface functions have been created to compensate for these differences, but several of them have failed in the regions studied. The main reasons for this are a lack of transparency in the transfer process (especially for SMEs, which are unaware of the transfer possibilities and/or may have reservations about utilising them), weaknesses in co-ordination (the services of different institutions often overlap), a failure of technology transfer infrastructure to meet the needs of the target groups (they may be supply driven) and the failure to concentrate regional strategies on the specific problems and economic potentials of the area. The existence of innovation support organisations alone does not guarantee networking, collective learning and knowledge development. For example, technology fairs provide networking and technology transfer opportunities but they only have a strong impact on high-technology SMEs in Milan; in five regions they are of little importance.

An interim result of our analysis was that the research and educational infrastructure of the regions studied was a necessary, although not sufficient, determinant of high-technology growth. On the other hand, the function of transfer institutions, technology-oriented fairs and innovation centres appears to be highly overrated. Of greater than expected importance were 'key people'. One such notable individual was F.J. Strauß in his role as Defence Minister and Head of the Federal German State of Bavaria. He used this to influence the positioning of military and R&D facilities in the Munich region. Key figures also exhibited decisive influence in the establishment of the technology park in Sophia-Antipolis (the current senator of the Département, Pierre Laffitte) and Oxford (Sir Martin Wood, founder of Oxford Instruments, established the Oxford Trust).

In addition, large firms can also influence the creation and long-term prosperity of an innovative milieu. An important determinant of this is the extent to which large firms participate in regional networking. One important example is Siemens with a predominantly positive role for Munich's high-technology sector (see Sternberg and Tamásy, 1999). It appears to be sufficiently embedded in the region to keep up locational ties with local R&D-intensive activities (although not for production) in spite of the influence of globalisation forces. Dicken, Forsgen and Malmberg (1994) point out two major influences for transnational corporations: first, parent company policy towards the outsourcing of functions and, second, the extent to which an environment meets their perceived requirements. The dependence on external decision centres constitutes an essential weakness historically characterising Sophia-Antipolis. Longhi and Quéré (1997) stress the difficulties of initiating local collective learning, when company strategies are decided outside the geographical area. Thus, the authors define Sophia-Antipolis as a 'very shaky innovative milieu that appeared progressively and is still developing in a very unstable environment' (Longhi and Quéré, 1997, p. 236). This situation however appears now to be changing (Longhi, 1999).

All in all, there is no clear-cut measure to determine the value and significance of institutional density. Regional innovation capacity depends on whether the individual elements of the system are compatible and work together. On a regional level, technology transfer and knowledge infrastructure can be built in such a manner that a high degree of synergy is generated. This means that, apart from the existence of institutions through which qualified employees, scientific findings and technology transfer are provided, the internal coherence and compatibility of regional frameworks must be taken into consideration (Braczyk and Heidenreich, 1998). In this respect, a successful example is Grenoble with its innovative milieu and distinctive networks involving public research, educational institutions (universities and technical institutes) and industry (de Bernardy, 1999). To the economic advantage of the region, the potential of technology is primarily established and efficiently used by endogenous processes rather than relying on politically motivated institutions for funding technology transfers.

3.5 Role of Technology Policy

How has technology policy shaped high-technology processes in the regions studied? In political terms, the answer to this question seems to lie in the dependency between the national, regional and local level (see section 3.2). It also depends on the interdependency between government influences and the other factors determining regional high-technology development including the supply of productive factors (e.g., a qualified workforce, risk capital), production networks and local demand conditions. In the regions studied, centres for knowledge creation (public and private research institutes; see section 3.3 and Appendix 3.1) exist, towards which national technology policy is oriented without following explicit regional goals (see Figure 3.3 for government R&D expenditure categorised in regions). From this technology base, players in each region acquire competitive advantages which ease the businesses' steps into new markets and raise the chances for R&D facilities to obtain additional state-funded R&D projects in the future. The implicit spatially-effective policy of nation states is only one link in the chain of cumulative regional development processes, which build on existing innovative potential dependent upon the type of funding (e.g. defence, civil; see Table 3.2). The importance of the armament policy in the United Kingdom is quite obvious, from which, however, the regions in the south east and south west of the country profit but not Cambridge and Oxford. For the Munich region, the arms industry has been particularly important for its development, which resulted from the strong military presence at this location (industries, research institutes) and extensive state funding from the German Defence Ministry.

The best example for national technology policy with explicit regional goals is Sophia-Antipolis, which arose almost entirely from measures adopted by the national government in co-operation with the Conseil Régional, the Départment and local authorities. Fifteen per cent of the

Table 3.2 Government budget appropriations or outlays for R&D categorised by socio-economic objectives (1996)

	Defence as percentage of total R&D budget	Economic development	Percentages of civil[1] R&D budget			General university funds
			Health and environment	Space	Non-specified	
Finland[2]	2.0	43.2	14.8	3.0	10.9	28.1
France[2]	29.0	19.1	12.5	15.3	27.0	22.5
Germany	9.8	23.1	12.7	5.5	16.5	41.3
Italy[3]	4.7	15.8	16.2	9.1	8.4	47.0
Netherlands	3.2	25.7	8.5	4.3	12.1	43.8
Spain	10.8	31.0	13.6	8.2	10.4	34.9
Sweden[1,2]	20.9	20.5	13.7	1.8	14.6	49.4
United Kingdom	37.0	16.6	31.7	4.3	18.3	28.4

Notes

1 For some countries, the categories do not add up to 100, because of residual categories.
2 Change in methodology.
3 1995.

Source: OECD 1998.

investment up to 1992 came from direct subsidies of the federal government and the Conseil Régional. Without going into great detail about the history of the region (Longhi and Masbouni, 1994; Longhi and Quére, 1997) one can note that federal intervention (e.g., the location of Air France and France Télécom, the re-location of public R&D facilities from the nation's capital) initiated the take-off phase at the end of the 1970s. The ball was got rolling by the transference of the elite Ecole Nationale Supérieure des Mines followed by the stimulus of the decentralisation plan of Mitterand in 1982. Since 1988 the technology park has been managed by a financing and development society, which received its mandate from the development syndicat (SYMIVAL: Syndicat Mixte pour l'Aménagement du Plateau de Valbonne) the members of which were from the local communities of Biot, Vallauris, Mougins, Valbonne and Antibes, the Départment Alpes-Maritimes, the Chamber of Industry and Commerce and the Chamber of Agriculture. In the past a shifting of political concentration from the national to the regional level has, thus, taken place (Charbit et al., 1991).

It is a salient feature of the last 20 years that regional authorities have become increasingly engaged in technology policy, which is mainly the case in regionalised or federal states like Germany or Spain. But also in more centralised countries such as France or Finland the promotion of new technologies at the regional level is viewed as a major component of economic development (European Commission, 1997). One successful example is Grenoble, where explicit and extensively consensual technology policy has been followed for a long time on the regional and local levels (Charbit et al., 1991). A symbol of these efforts was the establishment of ZIRST (see Appendix 3.1 for an explanation of the abbreviation) in 1972. Parallel to these measures public R&D facilities (such as CENG, CNET) and high-technology businesses outside the region were acquired, partly as a response to the developing innovative milieu of ZIRST (de Bernardy and Loinger, 1997). Technology policy in Grenoble is characteristic and exemplary in the sense of concerted action of all significant local players in the regional innovation system in concert with the goals of the federal government. However, with respect to the importance of regional technology policy Grenoble is the exception rather than the rule.

All in all, we can distinguish five types of technology regions (see Figure 3.4). Sophia-Antipolis, is a *'state-led high-technology complex'* resulting from technology policy with an intended regional goal. This has propagated a technology-oriented growth centre in the form of a 'new industrial space' (Scott, 1988). By contrast, *'state-facilitated high-technology complexes'*

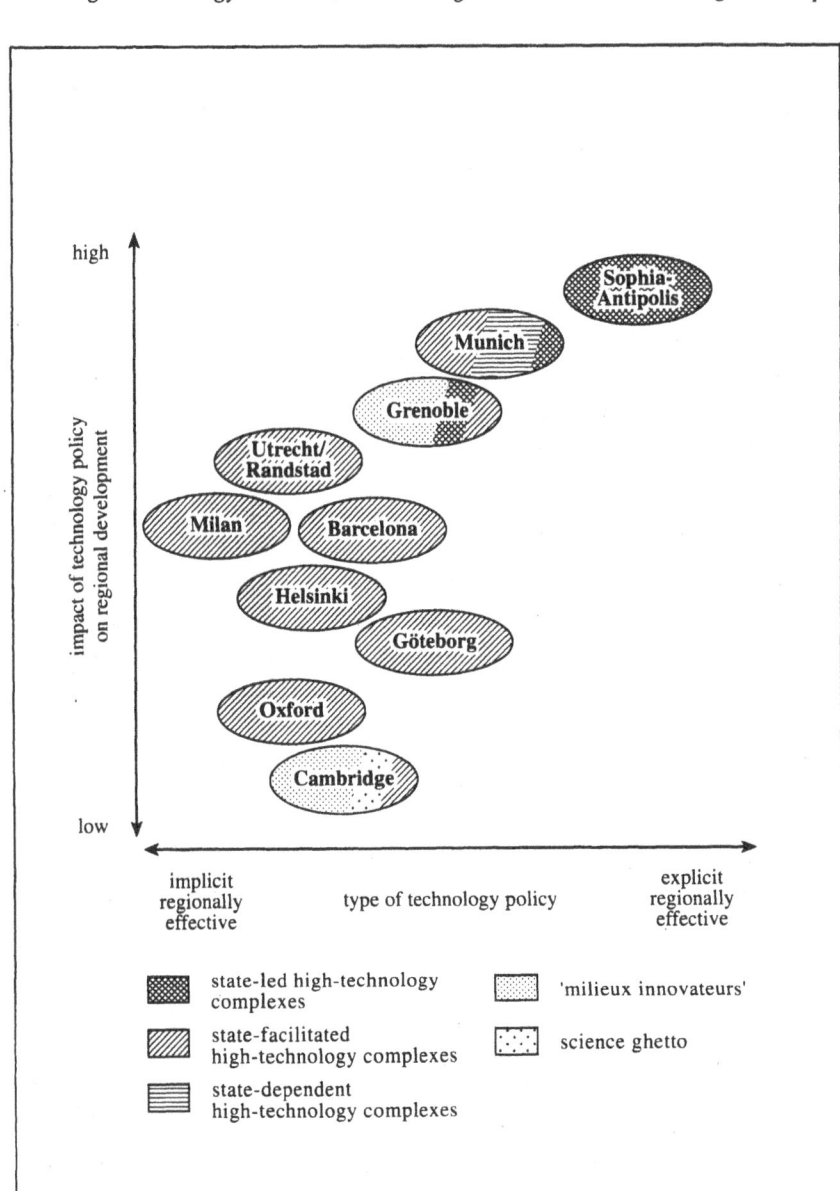

Figure 3.4 European regional clusters of high-technology SMEs: towards a typology

Source: own investigation, following the typologies of Gordon (1991) and Sternberg (1996).

attempt to utilise comparative advantages in the regions, but they are supported by central and regional governments primarily through policies with an implicit regional effect. The majority of the regions studied possess such characteristics (Barcelona, Göteborg, Helsinki, Milan, Utrecht/Randstad). The Munich region shares these characteristics but these are enhanced by important features of *'state-dependent high-technology complexes'* – characterised by massive public investments and R&D expenditures in the military area. Cambridge and Oxford are special cases which receive little support from technology policy, either explicit or implicit. Cambridge can perhaps best be described as a mixture of *'science ghetto'* and *'milieu innovateur'*. The significance of the university is proverbial, but the region left the ivory tower ghetto years ago. The close ties between the dominant SMEs and the university are important elements, which characterise innovative milieux and help them survive. Grenoble can be placed comparatively easily in the category of the *innovative milieu*.

3.6 Conclusions

Networking, collective learning and knowledge development in regionally-clustered high-technology SMEs in Europe can be significantly influenced by regional and local level policy measures. On the other hand, funding is more available nationally than regionally. Therefore: 'Regional innovation policies cannot replace national policies for techno-industrial innovation' (Hilpert, 1991, p. 21). But the region is the best source of knowledge about the demands on region-specific technology policy designed to improve regional innovation systems, especially for SMEs. As a consequence, funds can be (potentially) more efficiently utilised by regional than national technology policy. There would therefore appear to be a need to co-ordinate national, regional and local policy. However, with the exception of Grenoble, concerted policy making by policymakers at each level is uncommon and is hindered by a lack of local autonomy in Britain. In various regions, (for example, Helsinki, Milan, parts of the Randstad) attempts have been made recently which will possibly be successful in the mid-term. In any case, a closer co-ordination of regional initiatives with EU and national policy is required if regional competence centres are to fit in with wider policy objectives. At the same time, the transparency of the system of multiple direct and indirect programs for funding R&D projects must be increased especially if they are to benefit SMEs.

However, policymakers should avoid duplicating policies that have been successful elsewhere without a preliminary critical analysis of their suitability in different regional circumstances. Technology transfer infrastructures need to be shaped to meet the needs of particular regional economic and industrial structures so as to remove supply-side barriers to the outflow of usable new knowledge from universities and other public research institutions. From the demand side, capabilities and motivation of SMEs to use knowledge spill-over are essential. An important part of policy, therefore, is the overcoming of the obstacles to effective relationships between academic institutions and business and the development of trust and mutual confidence. It follows that if they are to be effective, the institutional and policy framework in any region must be industry as well as place-specific (Malecki, 1997) and this limits the identification of a generally applicable 'best practice' model.

Notes

1. Similar arguments are made by Amin and Thrift (1994) in their multifaceted concept of 'institutional thickness'.
2. We would like to thank all research network members for their support and for providing such an abundance of information about their regions.
3. The difference between the best and worst regions.
4. No regional data is available for Sweden.
5. The quantitative and qualitative evaluation of technology transfer and knowledge infrastructure is based on literary studies (just like the following section on the role of technology policy), research via the internet (self-portrayal of institutions, information from the Exhibition and Trade Fair Committee of the German Economy (AUMA): http://www.auma.de), and details from research network members in the regions to whom the same catalogue for evaluation was presented – see Table 3.1 and Appendix 3.1). The reports developed in the framework of the TSER network serve as a foundation (see Keeble and Lawson, 1996, 1997a, 1997b, 1998a, 1998b).

References

Amin, A. and Pywell, C. (1989), 'Is Technology Policy Enough for Local Economic Revitalization? The Case of Tyne and Wear in the North East of England', *Regional Studies*, vol. 23, pp. 463–77.

Amin, A. and Thrift, N. (1994), 'Living in the Global', in A. Amin and N. Thrift (eds), *Globalization, Institutions, and Regional Development in Europe*, Oxford University Press, Oxford, pp. 1–22.

Regional Institutional and Policy Frameworks 73

Bernardy, M. de and Loinger, G. (1997), 'Technopolis and Serendipity. The Adjustment Effects Created by Nancy-Brabois and ZIRST Technological Parks: A Comparison', in R. Ratti, A. Bramanti and R. Gordon (eds), *The Dynamics of Innovative Regions*, Ashgate, Aldershot, pp. 237–63.

Braczyk, H.-J. and Heidenreich, M. (1998), 'Regional Governance Structures in a Globalized World', in H.-J Braczyk, P. Cooke and M. Heidenreich (eds), *Regional Innovation Systems*, UCL Press, London, pp. 414–40.

Breschi, S. and Malerba, F. (1997), 'Sectoral Systems of Innovation: Technological Regimes, Schumpeterian Dynamics, and Spatial Boundaries', in E. Edquist (ed.), *Systems of Innovation: Technologies, Institutions and Organisations*, Pinter, London, pp. 130-156.

Charbit, C., Gaffard, J.L., Longhi, C., Perrin, J.C., Quéré, M. and Ravix, J.L. (1991), 'Modes of Usage and Diffusion of New Technologies and New Knowledge', *Fast Occasional Paper* No. 235, Commission of the European Communities, Brussels.

Cooke, P. (1998), 'Introduction: Origins of the Concept', in H.-J. Braczyk, P. Cooke and M. Heidenreich (eds), *Regional Innovation Systems: The Role of Governances in a Globalized World*, UCL Press, London, pp. 2–25.

de Bernardy, M. (1999), 'Reactive and Proactive Local Territory: Co-operation and Community in Grenoble', *Regional Studies*, vol. 33, pp. 343–52.

Dicken, P., Forsgen, M. and Malmberg, A. (1994), 'The Local Embeddedness of Transnational Corporations', in A. Amin and N. Thrift (eds), *Globalization, Institutions, and Regional Development in Europe*, Oxford University Press, Oxford, pp. 23–45.

Edquist, C. (1997), 'Introduction. Systems of Innovation Approaches - Their Emergence and Characteristics', in E. Edquist (ed.), *Systems of Innovation: Technologies, Institutions and Organisations*, Pinter, London, pp. 1–35.

European Commission (1997), *Second European Report on S&T Indicators 1997*, Brussels, Luxembourg.

Eurostat (1997), *Research and Development: Annual Statistics 1997*, Luxembourg.

Garnsey, E. (1998), 'The Genesis of the High Technology Milieu: A Study in Complexity', *International Journal of Urban and Regional Research*, vol. 22, pp. 361–77.

Gordon, R. (1991), 'Innovation, Industrial Networks and High-Technology Regions', in R. Camagni (ed.), *Innovation Networks: Spatial Perspectives*, Belhaven Press, London, pp. 174–94.

Hilpert, U. (ed.) (1991), *Regional Innovation and Decentralization: High Tech Industry and Government Policy*, Routledge, London.

Keeble, D. and Lawson, C. (eds) (1996), *Regional Institutional and Policy Frameworks for High-Technology SMEs in Europe*, ESRC Centre for Business Research, University of Cambridge, Cambridge.

Keeble, D. and Lawson, C. (eds) (1997a), *University Research Links and Spin-Offs in the Evolution of Regional Clusters of High-Technology SMEs in Europe*, ESRC Centre for Business Research, University of Cambridge, Cambridge.

Keeble, D. and Lawson, C. (eds) (1997b), *Networks, Links and Large Firm Impacts on the Evolution of Regional Clusters of High-Technology SMEs in Europe*, ESRC Centre for Business Research, University of Cambridge, Cambridge.

Keeble, D. and Lawson, C. (eds) (1998a), *Collective Learning and Knowledge Development in the Evolution of Regional Clusters of High-Technology SMEs in Europe*, ESRC Centre for Business Research, University of Cambridge, Cambridge.

Keeble, D. and Lawson, C. (eds) (1998b), *Regional Reports, ESRC Centre for Business Research*, University of Cambridge, Cambridge.

Kline, S.J. and Rosenberg, N. (1986), 'An Overview of Innovation', in R. Landau and N. Rosenberg (eds), *The Positive Sum Strategy: Harnessing Technology for Economic Growth*, National Academy Press, Washington D.C., pp. 275–306.

Lawton Smith, H. (1998), 'Barriers to Technology Transfer: Local Impediments in Oxfordshire', *Environment and Planning C: Government and Policy*, vol. 16, pp. 433–48.

Lindholm Dahlstrand, A. (1997), 'Entrepreneurial Spin-off Enterprises in Gothenburg, Sweden', *European Planning Studies*, vol. 5, pp. 659–73.

Longhi, C. (1999), 'Networks, Collective Learning and Technology Development in Innovative High Technology Regions: The Case of Sophia-Antipolis', *Regional Studies*, vol. 33, pp. 333–42.

Longhi, C. and Masboungi, J. (1994), 'Sophia-Antipolis: Historie et Devenir', paper for the Symposium Européen de Recherche sur les Technopoles, Rennes 5–7 April 1994.

Longhi, C. and Quéré, M. (1997), 'The Sophia-Antipolis Project or the Uncertain Creation of an Innovative Milieu', in R. Ratti, A. Bramanti and R. Gordon (eds), *The Dynamics of Innovative Regions*, Ashgate, Aldershot, pp. 219–236.

Malecki, E.J. (1997), *Technology and Economic Development: The Dynamics of Local, Regional and National Competitiveness*, Longman, Harlow (2nd edn).

National Board for Industrial and Technology Development (NUTEK) (1997), *Swedish Industry and Industrial Policy 1997*, Stockholm.

Nelson, R.R. (ed.) (1993), *National Innovation Systems: A Comparative Analysis*, Oxford University Press, Oxford.

OECD (Organization for Economic Co-operation and Development) (1998), *Science, Technology and Industry Outlook 1998*, Paris.

Scott, A.J. (1988), *New Industrial Spaces: Flexible Production Organization and Regional Development in North America and Western Europe*, Pion, London.

Sternberg, R. (1996), 'Government R&D Expenditure and Space: Empirical Evidence from Five Industrialized Countries', *Research Policy*, vol. 25, pp. 741–58.

Sternberg, R. and Tamásy, C. (1999), 'Munich as Germany's No. 1 High-Tech Region: Empirical Evidence, Theoretical Explanations and the Role of Small Firm/Large Firm Relationships', *Regional Studies*, vol. 33, pp. 367–77.

Stoneman, P. (1987), *The Economic Analysis of Technology Policy*, Clarendon Press, Oxford.

Tamásy, C. (1999), 'Evaluation of Innovation Centres in Germany', paper for the Babson-Kauffmann Entrepreneurship Research Conference, University of South Carolina, 12–15 May, Columbia.

Appendix 3.1 Selected characteristics of regional innovation systems

BARCELONA
Six **universities** with a total of more than 160,000 students: *University of Barcelona*: 74,000 students, 3,000 faculty members, technologically relevant fields of study: physics, chemistry, electrical engineering, responsible for technology transfer: Fundació Bosch i Gimpera (est. 1983); *Universitat Autònoma de Barcelona*, 1996: 36,856 students, 2,643 faculty members, technological relevant fields of study: mathematics, computer sciences, physics, chemistry, biology, genetics, micro-biology, pharmacy, food technology, technology transfer office: Officina d'Investigació i de Transferència de Tecnologia (OITT), a few research centres and affiliated institutes (like the Institute of High Energy Physics); *Pompeu Fabra University*: 7,000 students, social sciences; *Universitat Oberta de Catalunya*: open distance-learning university based on the concept of a virtual campus; *Universitat Ramon Llull*, 1997/98: 12,543 students, 633 lecturers, two offices for technology transfer: PEINUSA (field of chemical industry, company services), NexTReT (field of electronics, telecommunications, computer sciences and multimedia); *Universitat Politècnica de Catalunya*: approx. 30,000 students, 3,000 professionals, technologically relevant fields of study: mathematics, industrial engineering, telecommunication engineering, computer sciences, CTT (Technology Transfer Centre, est. 1987), research centre: IRI (Institute of Industrial Robotics and Informatics), several laboratories. **Outside the university, (partially-) state-owned R&D facilities**: *LGAI* (best equipped research and test laboratory in the region); *ABC* (Catalan Agency for Biotechnology); *IRTA* (agro-alimentary research institute); basic research mainly carried out by: *CSIC* (National Public Research Centre), 11 CSIC centres in Catalonia, three on the campus of the Universitad Autònoma de Barcelona. **Technology parks/innovation centres**: *Barcelona Science Park*: in the city of Barcelona; in phase one (1996–2000) approx. 23,000 m^2 of space for experimental research and offices available, up to 2008 extension planned up to 56,000 m^2; three kinds of tenants: academic institutes (mainly in the field of biomedical

research), mixed research institutes (private and university owned), companies (e.g., Merck; installing a molecular biology laboratory);
Parque Tecnológico des Vallés: approx. 30 km west of Barcelona, 60 ha, for 120 facilities, founded in 1988 neighbouring subsidiaries of multinational businesses (e.g., Hewlett Packard, Sharp).
Fairs:
HISPACK: International Exhibition for Bottling and Packaging, 1995: 1,557 exhibitors, 27,069 visitors; *EXPOTRONICA: International Trade Fair for Electronic Equipment and Components* (electrical engineering, electronics, c-technology, automated production, measurement, control and regulation engineering), 1996: 656 exhibitors – 501 from foreign countries, 8,140 visitors; *SONIMAG: International Exhibition for Televisions, Radios and Electronics* (electronic media, radio and television technology), 1996: 487 exhibitors – 328 from foreign countries, 30,943 visitors; *INFORMAT: International Exhibition for Computer Science* (computer hardware, software, personal computer, telecommunications), 1996: 418 exhibitors – 219 from foreign countries, 25,600 visitors.
Additional Information:
CIDEM (Business Information and Development Centre, co-operation between research institutes and industry); *CIRIT* (facility responsible for carrying out Catalan technology policy); *Metropolitan Agency for Urban Development and Infrastructure* (area projects: airport enlargement, building the Llobregat Delta Logistic Area near the airport and harbour, port enlargement, new high-speed train station); 74% of those employed work in the service sector.

CAMBRIDGE

Two **universities** with a total of approx. 22,000 students:
University of Cambridge, 1998: approx. 15,800 students, 1,400 teaching staff and 2,000 other research staff, world-wide reputation in technologically relevant fields of study: biology, chemistry, computer science, genetics, manufacturing engineering, materials science, pharmacology, physics, since 1996 new Institute of Biotechnology;
Anglia Polytechnic University, 1998: 6,200 students, 300 academic staff, technologically relevant fields of study: information systems, biological sciences, biomedical science, chemistry, computer science, electronics, engineering.

Outside the university, (partially-) state-owned R&D facilities:
Sanger Centre (human genome project, at Hinxton Hall); *CAD Centre* (Computer Aided Design Centre); institutes in the fields of: biotechnology (e.g., *MRC Laboratory for Molecular Biology*, *Babraham Institute of Animal Physiology*, *NIAB* (National Institute of Agricultural Botany), *PBI* (Plant Breeding International), *Dunn Nutrition Unit*) and material sciences.

Technology parks/innovation centres:
Cambridge Science Park (est. 1970), 75 high-tech companies, 4,000 employees; *Granta Park* (11 km south of Cambridge): approx. 8,000 m^2 office space (in 2000); *St. John's Innovation Centre* (est. 1987), incubator for technology-oriented businesses, about 60 companies, 1,000 employees, close to 90% of the companies survived after outgrowing the science park (1987–98), also running programmes for firms outside the innovation centre; additional science parks in Cambridge: *Melbourn Science* Park (15 km south of Cambridge), Cambridge Research Park (just north of Cambridge), and planned biotechnology incubator park at Hinxton Hall (south of Cambridge); various other business parks, some of them containing several high-technology firms (e.g., *Vision Park*, *Castle Park*).

Fairs: no relevant technology fair.

Additional Information:
Rapid local growth of high-technology industry (Cambridge Phenomenon); approx. 27,000 people employed in about 800 high-technology companies in and around Cambridge (1998), of these about 250 businesses in computer hardware, software and services, 23 in telecommunications; four research consulting firms of world-wide reputation (*Cambridge Consultants, Scientific Generics, PA Technology, The Technology Partnership*); *Industrial Liaison and Technology Office* (ILTO, formerly Wolfson Industrial Unit), technology transfer between University of Cambridge, research institutes and industry; *Cambridge Research and Development Ltd.* (CRIL), focused on developing businesses based on newly emerging technologies; *Cambridge Quantum Fund* (CQF), supports technologies emanating from the University; *Business Link Central and Southern Cambridgeshire* (business support organisation); wide range of Business Clubs and other opportunities for informal discussion, e.g.: *Cambridge Europe and Technology Club*, *Cambridge High-Tech Association of Small Enterprises* (CHASE),

Cambridge University Local Industry Links (CULIL); several venture capital funds.

GÖTEBORG
Two **universities** with a total of approx. 42,300 students:
Chalmers University of Technology, 1996/97: approx. 8,300 students, 998 doctoral students, 1,427 faculty members and researchers, approx. 1,000 ongoing research projects, 40% of Sweden's graduate engineers and architects were educated at Chalmers, fields of study: automation engineering, bio-engineering, computer science, electrical engineering, engineering physics, chemical engineering, research in the fields of: mathematics, natural sciences, engineering, industrial sciences, microelectronics, energy and material science (partially in co-operation with Göteborg University);
Göteborg University, 1997: approx. 34,000 students, 1,900 postgraduate students, 2,300 faculty members/researchers, 250 professors, six faculties (arts, social sciences, medicine, odontology, science, applied and fine arts).
Outside the university, (partially-)state-owned R&D facilities:
Seven independent industrial research institutes, main fields of research: production engineering, biochemistry, polymers, ceramics, etc., including: *The Swedish Institute for Food and Biotechnology* (SIK), *Swedish Institute for Fibre and Polymer Research* (IFP); *Chalmers Contract Research Organisation*; six *National Competence Centres* work as an interface between Chalmers University and the industry (Chalmers University, companies and the NUTEK (National Board for Industrial and Technical Development, co-ordinating Sweden's participation in EUREKA) hold a $^1/_3$ share in the Competence Centres).
Technology parks/innovation centres:
Chalmers Science Park (est. 1987) in the grounds of Chalmers University of Technology, 10,000 m^2 of laboratory and office facilities for industrial research for approx. 350 people, high-technology and research-based companies can set up long-term research and development operations in collaboration with Chalmers University; Chalmers Innovation Centre (CIC) (construction decision 1996, Sweden's largest science park investment, 5,600 m^2 floor space planned); Stena Centre north of Chalmers University with facilities for spin-off companies; Chalmers House of Innovations (incubator, on the Chalmers University campus, 1998: seven spin-offs).

Fairs:
KOMPONENT: Trade Fair for Electronic Parts (electronic parts, measuring technology, test equipment), 1997: 2,834 exhibitors, 9,026 visitors; *SCANAUTOMATIC:* International Fair for Hydraulics, Pneumatics, Electronics, Transmission and Control Engineering, 1997: 2,277 exhibitors, 14,914 visitors; *ELFACK/ ELKRAFT:* International Fair for Electrical Engineering, Energy Production and Distribution, 1997: 1,230 exhibitors, 25,609 visitors; *MILJÖTEKNIK/ ECO-TECH:* Trade Fair for Environmental Technology, 1997: 533 exhibitors, 9,873 visitors; *KEMI:* International Trade Fair for Chemistry (analysis technology, chemical products, laboratory facilities), 1995: 227 exhibitors, 6,059 visitors; *TELECOM SCANDINAVIA NETWORK EXPO* (computer technology, telecommunications), 1998: 108 exhibitors, 5,386 visitors.
Additional Information:
Approx. 240 direct university spin-off companies from Chalmers University between 1960 and 1993; *Industrial Liaison and Development Office* (approx. 70 research and industrial projects each year, enhance collaboration between University and Industry); *Chalmers Innovation Centre* (CIC) supports innovation activities.

GRENOBLE
Four **universities** with a total of approx. 50,000 students:
Université Pierre Mendes France, 1998: approx. 20,000 students, approx. 500 faculty members, 680 researchers, 32 groups of researchers of which 24 have an official reputation/recognition, fields of study: social sciences; *Université Joseph Fourier*: approx. 18,000 students of science, technology, medicine, (attached) institutes: Centre for Innovations in Multimedia, Institute of Computer Science and applied Mathematics, Institute of Sciences et Techniques, Institute of Technology 1 (IUT 1), Laboratory of Spectrometric Physics, Medical/Pharmaceutical Research Institute;
Université Stendhal, 1997/98: 8,000 students, 320 faculty members, 19 research centres, five fields of study (language sciences, communications); *Institut National Polytechnique de Grenoble* (INPG): technological university, 4,300 students, 1,250 faculty members (345 full-time), 1,400 researchers, 30 laboratories, 25 associated to the CNRS (fields of research: computer science, applied mathematics, mechanical engineering, materials and process engineering, automation, micro-

electronics, nine engineering schools (concentrations: electronic chemistry, electronics, hydraulics, applied mathematics, physics, industrial systems);
Two technical colleges with a total of 3,600 students.
Outside the university, (partially-) state-owned R&D facilities:
220 laboratories with approx. 17,000 research personnel, 13,000 in public research, 4,000 in private research (10,000 in fundamental research, 7,000 in applied research), research facilities of five national research centres: *CEA* (Atomic Energy; also under the CEA Directorate of Advanced Technologies: *LETI* (information technology, micro-technologies, serving the industry, 936 employees), *CNET* (National Centre for Telecommunication Studies), *CNRS* (Centre National de la Recherche Scientifique), *CRSSA* (medicine), *INRIA* (information technology and automation, close ties with *Imag* (Institute for Applied Mathematics and Computer Science); international institutions, like: *Pasteur Institute, CIRC* (cancer research), *ESRF* (basic and applied research in physics, chemistry, materials and life sciences, 500 employees), *ILL* (fundamental research institute, operating a high-flux reactor, 80 scientists, 300 other staff), *IRAM, GHMFL, EMBI, The Open Group – Grenoble Research Institute* (research on fundamental software technology); several *private research laboratories*; region-specific institutions, like: *CENG* (centre for nuclear studies; 2,500 employees).
Technology parks/innovation centres:
ZIRST (Zone for Realisation, Scientific and Technological Productions; est. 1972) in the suburbs (municipalities of Meylan and Montbonnot), location for high-tech firms, 1998: 230 firms on 110 ha, 5,330 employees; about $^2/_3$ of the firms have less than 10 employees.
Fairs:
TEC: European Forum for Competitive Technology (data-processing facilities, electronics, materials, microelectronics), 1996: 370 exhibitors – 115 from foreign countries, 12,259 visitors.
Additional Information:
Approx. 65% of the jobs in the Grenoble/Isere region are in the service sector; second most important location in France for strategic jobs, with 23,000 people in information and communication technology, public and industrial research, industrial marketing, management and services; important European centre for microelectronics; settlement of approx. 150 international large corporations in the Isere region of their own

accord; 6,000 jobs in bio-medical research and industry; *Grenoble Network Initiative* (GNI): promote/develop telecommunication infrastructure.

HELSINKI

Two **universities** with a total of approx. 46,000 students in the 'Helsinki Metropolitan Area':
University of Helsinki, 1997: 32,800 students, approx. 2,500 scientists in research and teaching, of these 479 are professors or associate professors, 1995: 40% of the Finnish doctor theses, 9 faculties, numerous attached institutes (e.g.: Computer Centre, Institute of Biotechnology, Research Institute for High Energy Physics);
Helsinki University of Technology (in Espoo), approx. 13,000 students, 181 professors, fields of study: CAD, digital image processing, semiconductor technology, automation and robotics, surface technology, material development, medicine and energy technology;
Helsinki School of Economics and Business Administration, 3,800 students; *Technical Institute of Helsinki*, 2,300 students, 250 faculty members; three polytechnic institutes in the Helsinki region: *Technical Institute of Helsinki*, approx. 2,500 students, *EVItech*, 2,000 students in engineering degree programs: automation technology, biotechnology, computer engineering, information technology, medicine technology, etc.
Outside the university, (partially-) state-owned R&D facilities:
The Academy of Finland (purpose: advance scientific research, promote international research co-operation); *Finnish Meteorological Institute*; *Finnish Centre for Radiation and Nuclear Safety*; *National Veterinary and Food Research Institute* (e.g.: bacteriology, food microbiology, food hygiene); *Technical Research Centre of Finland VTT*, in Espoo, 2,700 employees, building technology, information technology, chemical technology, mechanical technology, important contracting research centre for the industry and the public sector.
Technology parks/innovation centres:
Helsinki Science Park (est. 1993) on the grounds of the Faculty of Agriculture and Forestry (University of Helsinki) 10 km north of the Helsinki town centre, focus on biotechnology, together with the faculty it forms the Viikki Green Valley, a centre for bio-sciences; *Otaniemi Science Park* in Espoo near the Helsinki University of Technology, largest centre for technological knowledge and business in Northern

Europe, located in the park: headquarter of the VTT (Technical Research Centre of Finland), Innopoly Technology Centre and Otech (centre for new companies in information technology and telecommunications), about 200 companies, approx. 1,050 employees, focuses on: information technology and telecommunications; *Biomedicum*, in developmental stage, bio-medical science park, next to the University Central Hospital in Helsinki; *Airport City*, in developmental stage, centre for technology-based companies, near the Helsinki-Vantaa International Airport; *Business and Service Park Spektri* (near the Otaniemi Science Park), location of multinational high-technology companies.
Fairs:
ELKOM-ELTEC: International Fair for Trade Electronics (CAD/CAM, data transfer, electronic equipment, parts and systems, energy generation, energy technology, energy distribution, materials, measurement technology, test equipment), 1997: 2,099 exhibitors, 24,888 visitors; *FinnTec: International Technology Fair* (automation, CAD/CAM, industry robots, maintenance, measuring equipment and technology), 1997: 1,702 exhibitors, 33,703 visitors; *AUTOMATION: Trade Show for Automation* (c-technology, production automation, measurement, control and regulation technology), 1997: 1,160 exhibitors, 20,919 visitors; *Tt-Office Technology-Fair* (data processing, telecommunications), 1997: 611 exhibitors, 25,505 visitors; *HOME COMPUTERS/Voice & Vision: Exhibition for Consumer Electronics*, 1997: 82 exhibitors, 19,525 visitors.

MILAN
Five **universities** with a total of approx. 182,500 students:
Universita degli Studi di Milano, 1997: approx. 96,600 students, 719 faculty members, 12 faculties, among them: Faculty of Economics, Mathematical, Physical and Natural Sciences, Medicine and Surgery, Pharmacy, inter-faculty degree course in biotechnology;
Universita Commerciale Luigi Bocconi (private), 1997: approx. 12,100 students, 670 faculty members;
Universita Cattolica del Sacro Cuore (private), 1997: approx. 37,780 students, 2,146 faculty members;
Milano Politecnico, 1987/88: 32,925 students, 740 scientists;
Milano IULM (private), 1987/88: 3,120 students, 66 scientists.
Outside the university, (partially-) state-owned R&D facilities:
Centre for Information, Studies and Experiments SpA (CISE); *Mario*

Negri; *ASSORENI* (chemistry, applied engineering); *CESI* (electrical engineering); *more institutes of the national research centre CNR*; *IRB* (new materials); more business-owned research centres.
Technology parks/innovation centres:
Technology and Environment Park (est. 1992) in Sesto San Giovanni near Milan.
Fairs:
SMAU: International Fair for Information and Communication Technology (c-technology, data processing systems, communication technology, network technology, computer hardware/software), 1997: 2,840 exhibitors – 893 from foreign countries, 356,656 visitors; *BIAS AUTOMAZIONE MICROELCTRONICA: International Exhibition for Automation, Instrumentation and Microelectronics* (production automation, c-technology, measurement, control and regulation technology), 1996: 2,560 exhibitors – 1,547 from foreign countries, 58,912 visitors; *MAC: Fair for chemical Analysis, Research and Control Equipment and Biotechnology*, 1997: 1,274 exhibitors – 676 from foreign countries, 26,469 visitors; *INTEL: International Trade Fair for Electrical Engineering and Electronics*, 1995: 1,172 exhibitors – 186 from foreign countries, 83,868 visitors; *IBTS: International Fair for Film, Radio and Telecommunications*, 1997: 459 exhibitors – 215 from foreign countries, 21,978 visitors; *MIFED: International Multi-media Market* (television, movie and theatre technology), 1996: 283 exhibitors – 254 from foreign countries, 4,310 visitors; *SICOF: International Trade Fair for Film, Photo, Optics, Audio-vision, Lab Technology*, 1997: 252 exhibitors – 109 from foreign countries, 31,186 visitors.

MUNICH

Ten **universities and polytechnic colleges** with a total of more than 103,000 students:
Ludwig Maximilian University (LMU), 1992/93: 63,585 students, 3,440 faculty members, 90 courses of study, fields of study: languages, law, economics and social sciences;
Technical University (TU), 1994/95: 19,991 students, 1994: total of 3,410 faculty members (of these 411 are professors), numerous international partnerships and double degrees, fields of study: mathematics, mechanical engineering, natural sciences;
Polytechnic College of Munich: 17,000 students, 480 professors, 15

technical fields of study (of 20 in total);
University of the Armed Forces Munich: 2,700 students, 200 faculty members, fields of study: engineering, computer sciences, air and space technology;
Polytechnic College Weihenstephan; *Ukrainian Free University*; *Academy for Film and Television*; *Academy for Politics*; *Academy for Music*; *Academy for Philosophy*.

Outside the university, (partially-) state-owned R&D facilities:
General administration of the *Max-Planck-Society* and 11 of its institutes (approx. 6,000 employees), among these: bio-chemistry, physics and astrophysics, plasma-physics, quantum-optics; central unit of the *Fraunhofer Society to Promote Applied Research* and three of its institutes (solid-state technology, food stuff technology and packaging); *GSF-Research Centre for Environment and Health, DLR German Research Society for Aerospace Travel*, leading centre for R&D in the private economy; *Paper Technology Foundation*; *Research Institute of Society for Radiation and Environmental Research*; *Max von Pettenkofer Institute for Hygiene and Medical Microbiology*.

Technology parks/innovation centres:
Munich Technology Centre (MTZ), 11,000 m^2 area for high-tech businesses, near the airport, 1994 approx. 60 businesses, managed by the Fraunhofer Management Association; *Technology Centre Rosenheim*; Innovation Centre for biotechnology in Martinsried (est. 1998), 5,600 m^2 of rental space.

Fairs:
Since 1998 new convention centre in Riem with 145,000 m^2 of space, important European location for international high-tech trade fairs with a total of approx. 350,000 branch visitors:
Electronica: International Trade Fair for Components and Electronic Parts (electronic components, semiconductor, circuit boards, microelectronics, microwave technology, optic electronics, sensory systems), 1996: 2,305 exhibitors – 1,110 from foreign countries, 78,677 visitors; *SYSTEMS: International Trade Fair for Information Technology and Telecommunications* (computer software, computer technology, data protection, information technology, communication technology, multimedia, network technology, telecommunications), 1997: 1,735 exhibitors – 147 from foreign countries, 107,563 visitors; *ANALYTICA: International Trade Fair for Analysis, Biotechnology, Diagnostic and*

Laboratory Technology, 1998: 991 exhibitors – 319 from foreign countries, 32,893 visitors; *LASER: International Trade Fair for Innovative and Applied Laser Technology and Optic-electronics*, 1997: 730 exhibitors – 345 from foreign countries, 14,121 visitors; *Transport: International Trade Fair for Logistics, Telematic, Freight Goods and Passenger Services*, 1997: 690 exhibitors – 227 from foreign countries, 27,558 visitors; *EXOPHARM: International Pharmaceutical Trade Fair* (laboratory instruments, pharmaceuticals), 1997: 490 exhibitors – 62 from foreign countries, 21,711 visitors; *INTER AIRPORT: International Trade Fair for Airport Technology and Services*, 1997: 486 exhibitors – 285 from foreign countries, 9,268 visitors; *ELTEC: Trade Fair for Electrical Engineering*, 1998: 480 exhibitors – 20 from foreign countries, 21,081 visitors.

Additional Information:
Businesses in Munich with outstanding R&D departments: Siemens (15,000 R&D employees), BMW (6,000 R&D employees), DASA; *Bavarian Academy of Sciences* (research funding, sponsor of various research ventures); 1995 about 111,000 persons were employed in the Munich region in high-tech companies; fields of study of high-tech firms: electronics, automobile manufacturing, aerospace industry; each university and each technical college maintains an agency responsible for technology transfer; inter-SME networks like the *Förderkreis Neue Technologien* (Sponsor for New Technologies) offers a forum for technology-intensive firms and science in Bavaria and the Munich region; funding in the region through the national technology policy (BioRegio Competition): up to November 1998 14 Projects were approved (= ten new firm establishments) with a total of 55 Mio. DM (Federal Ministry of Research and Technology's (BMBF) share of the funding: 27 Mio. DM); in the same time frame 30 new firm establishments in biotechnology in the region; strong presence of military facilities and firms at the same location (e.g., German Aerospace AG with MBB, Dornier and Telefunken Systemtechnik).

SOPHIA-ANTIPOLIS

One **university**:
Nice Sophia-Antipolis University, 1995/96: approx. 27,000 students, 1,150 faculty members, 200 subject and research centres;
ESSI (training in information technology); *CERAM* (industrial software

development, network management), 750 students, 100 full-time faculty members; *THESEUS Institute* (objective: bring together information technology and management); *ISIA* (telecommunications, automation, robotics); *National School for Advanced Engineering*, 350 students, 120 researchers/scientists.

Outside the university, (partially-) state-owned R&D facilities:
Facilities of the national R&D institutions: National Centre for Scientific Research (CNRS); *INRIA Sophia-Antipolis* (computer science research), located at Sophia-Antipolis Science Park, 400 employees, including 140 permanent staff; *Institut Eurécom* (research institute for training engineers in communication systems); 200 firms (Sophia-Antipolis share: 31%) with private research, 8,000 full-time researchers (Sophia-Antipolis share: 43%) in those firms.

Technology Parks/Innovation Centres:
Sophia-Antipolis Science Park: 1,050 companies (1995 including 97 foreign companies), 17,000 engineers and technicians, 5,000 researchers in an area of 5,750 acres, extension planned for 11,250 acres, main industries: information sciences, electronics, advanced telecommunications (22% of the corp.), health sciences, chemistry, biotechnology (5% of the corp.), environment, energy (2% of the corp.), 60 research institutions (7% of the corp.), approx. 2,500 students work in the science park; *Arénas International Business Centre* (Nice), close to the int. airport: 1,858,000 m^2 office space/facilities for high-tech companies; *Centre International de Communication Avancée* (CICA): 300 engineers, 300 students and researchers, 50 companies.

Fairs:
No national or international technology fairs, however, a few international events in the high-tech and medical domains (Acropolis Conference & Exhibition Centre).

Additional information:
83% tertiary activities, fields of research in the region: mathematics, information technology, electronics, telecommunications, physics, chemistry; International Nice-Riviera Airport; *Telecom Valley Association* (interaction between 40 companies in the field of telecommunications).

OXFORD

Two **universities** with a total of approx. 26,500 students:
University of Oxford: 15,500 students, 1,500 faculty members, of these 350 are professors, Faculty of Biological Sciences, Faculty of Clinical Medicine, Faculty of Physical Sciences (information engineering, materials science, mechanical engineering, chemistry, electrical engineering, electro-optic engineering, experimental physics, laser physics), receives heavy research investments from overseas especially in pharmaceuticals;
Oxford Brooks University: 8,274 students, (+ 2,749 part-time students), 700 faculty members, technologically relevant fields of study: engineering, biological and molecular sciences, computing and mathematical sciences, but no well-developed infrastructure for transferring technology.

Outside the university, (partially-) state-owned R&D facilities:
Seven national laboratories with a total of approx. 6,100 employees in 1997 in the county of Oxfordshire, including two belonging to the *UK Atomic Energy Authority* (UKAEA) (*Harwell and Culham Laboratories*) and the *Rutherford Appleton Laboratory* (approx. 25 km south of Oxford; research programs include: biology, chemistry, computing, engineering, materials science, space science).

Technology Parks/Innovation Centres:
The Oxford Science Park (est. 1991), associated with the universities of Oxford and other major centres of research, alliances with overseas science parks in the US, Japan and Australia, 40 science and technology-based companies in an area of 30 hectares, five km south of Oxford; technology business incubators: *Milton Park Innovation Centre* (in Abingdon, approx. 10 km south of Oxford), *Oxford Centre for Innovation* (OCFI), close to the city centre, run by The Oxford Trust (generating links between firms and institutions within the county), by now 29 early-stage technology based companies, *Cherwell Innovation Centre*, in the north of Oxfordshire.

Fairs: No relevant technology fair.

Additional information:
Oxfordshire Investment Opportunity Network (bring together entrepreneurs and investors); *Oxfordshire BiotechNet* (stimulate formation and growth of biotechnology and related companies in Oxfordshire); approx. 45 businesses in the field of biotechnology in Oxfordshire.

UTRECHT/RANDSTAD

Four **universities** with a total of more than 47,000 students:
Utrecht University, 1998: 22,047 students, 3,175 academic personnel, 21 research schools under the supervision of the university, participant in additional 30 research schools, whose fields of research include: biophysics, climate and environmental studies, health and health care (medical technology), e.g.: Centre for Bio-molecular Research, GIS Expertise Centre, Institute of Bio-membranes, Mathematical Research Institute; *Hogeschool van Utrecht*: 25,000 Students, 2,400 academic and non-academic staff, technically-relevant faculty: Faculty of Science and Engineering; *Utrecht University Hospital*, medical and pharmaceutical research carried out (electronic image processing, immunology, pulmonary diseases); *Katholieke Theologische Universiteit te Utrecht*, 390 students, 53 faculty members; *Universiteit voor Humanistiek*, 260 students, 26 faculty members.

Outside the university, (partially-) state-owned R&D facilities:
Foundation for the Fundamental Research of Matter (FOM, high-grade physics research); *Space Research Organisation Netherlands* (SRON); *Netherlands Institute for Developmental Biology* (NIOB); *Software Engineering Research Centre* (SERC); several expertise centres.

Technology parks/innovation centres: No relevant technology park or innovation centre.

Fairs:
Techni-Show: Trade Fair for Industrial Production Technology, 4,157 exhibitors – 3,436 from foreign countries, 65,238 visitors; *Elektrotechniek* (lighting technology, electrical engineering, electrical equipment, measuring tools, test equipment, control and regulation technology), 1997: 527 exhibitors – 40 from foreign countries, 39,858 visitors; *LOGISTICA: International Trade Show for Material Flow Technology*, 1997: 392 exhibitors – 38 from foreign countries, 29,663 visitors; *Industriele Electronica: Exhibition for Industrial Electronics* (electronic components, production facilities, control facilities, circuit boards, circuit board production, measuring instruments, programming systems for ROM, control and regulation technology), 1997: 271 exhibitors, 11,025 visitors; *ECOTECH EUROPE: International Trade Fair for Waste Removal and Environmental Technology*, 1997: 266 exhibitors – 33 from foreign countries, 20,441 visitors.

Additional information:
Province of Utrecht has a highly developed information and communication technology sector, approx. 6% of the total number of jobs in Utrecht (companies: AT&T, Cap Gemini, Digital, IBM, Sun Microsystems).

4 University and Public Research Institute Links with Regional High-Technology SMEs

HELEN LAWTON SMITH AND MICHEL DE BERNARDY

4.1 Introduction: Knowledge Institutions and Regional Economic Development

Universities and government research institutes (*knowledge institutions*) have been cast as lead players in a variety of economic strategies. Together with their traditional teaching and research responsibilities, they are now more accountable to governments, business and localities, each with their own changing sets of expectations. Because of its importance for industrial innovation, research by knowledge institutions is increasingly influenced by regulation and policy at the regional, national and international levels, by the business climate and by the attitudes of industry (Wever, 1997, p. 14). This chapter's concern is with the role of university and research institutes as regional collective agents and the evolution of their links with regional high-technology firms. It seeks to identify how knowledge institutions engage in different processes in different contexts within innovative milieu, and their places in local history as well as in national innovation systems. Our starting point is with the process of technology diffusion, which as originally conceived was a linear transfer of knowledge generated in research centres to industry. Technology transfer is now seen as a more interactive and geographically focused process resulting from the collective action of a variety of partners in which the direct transfer of technology from research centres plays a less central role.

Our understanding of how innovation works leads to a new paradigm in which activities are more oriented towards problem solving than the flow of new technology as in the pipeline model. This new paradigm implies a change of behaviour on the part of knowledge institutions towards more continuous innovation requiring more two-way research interaction with industry combined with a raft of other modes of contact.

In examining the roles which knowledge institutions play in their local economies we draw on a series of case studies which exemplify the different processes within their political contexts. We conclude with a comparison of Oxford and Grenoble. These are similar in that they both have concentrations of nuclear research laboratories and leading universities and are what Amin and Robins (1990) describe as 'pioneer regions'. They are different in that on the one hand, Oxford is in the process of developing its effective technology hinterland, while on the other Grenoble has been established as a centre for high-technology innovation since the beginning of the twentieth century.

4.2 Local Links and Milieu

Knowledge institutions are now recognised as being participants of 'the set of relationships occurring within a geographical area' (Camagni, 1993, p. 3) particularly within islands of innovation (EC, 1994, p. 203). While the nature of the local milieu is that of 'collective operator reducing the degree of static and dynamic uncertainty of firms', knowledge institutions are identified here as being collective agents whose multi-layered engagement with other local actors contributes to a place-specific cohesive set of relationships which facilitate collective learning. Their impact arises from being a source of highly skilled labour, of new firms and of technology. As incubators they influence the skill mix of the local labour markets through the supply of graduates and as a breeding ground for new entrepreneurs who become role models for others. They also train the existing local workforce through continuing education departments and attract highly qualified workers from outside the area (Cooper, 1971, p. 2). According to Camagni (1991), in the theory of milieu proximity to local sources of highly skilled labour which are highly mobile within the territory accounts for much of collective learning. As collective agents they therefore affect a region's capacity for technological development (Luger and Goldstein, 1991, p. 8). They have a further important milieu role of contributing to the cultural and psychological identity of localities.

However, the relationship between universities and small firms is not without problems as a recent study for the European Commission pointed out:

This 'link gap' between universities and the indigenous technological base is due to a clash of two quite different organisational cultures. At one end the traditional culture within many academic institutions does not encourage the development of links with small-scale industry. On the other hand, there is also evidence of reluctance by small technology-based firms within peripheral regions to become involved in relationships with their local universities although large companies regularly access university departments for external sources of technological expertise (Jones-Evans, 1998, p. 23).

While these barriers to links between universities and small firms refers to peripheral regions, they operate with equal force throughout Europe. They are based on a misunderstanding in universities about what small firms can provide in the way of funding and on the part of small firms of what universities can provide in the way of new technology.

This chapter is concerned with identifying, examining and evaluating the outcomes of processes by which relationships between knowledge centres and small firms are formed. It also examines the dominant factors which bring about changes. In this respect, Keeble and Lawson (1997, p.1) argue that linkages within an innovative milieu may well be particularly important during the early stages of development of small technology-intensive enterprises, many of which tend to be established as spin-offs from existing firms, universities or other institutions.

The discussion follows the structure set out in Table 4.1 which summarises ways by which regionally-based universities and public laboratories influence the process by which clusters of knowledge-intensive firms are formed and links with them are established and maintained. The context for this discussion is set in the next section by a description of the populations of the knowledge institutions in the case study regions and the kinds of institutional arrangements adopted.

4.3   Universities and National Laboratories in the Study Regions

The case study regions range in size and function from metropolitan regions (Barcelona, Milan, Göteborg, Helsinki and Munich); to smaller yet pivotal centres of public and private sector R&D (Oxford and Cambridge); to long established regions (Grenoble); to more diffuse systems of innovation embracing metropolitan and non-metropolitan regions (The Randstad); to a planned technopole (Sophia-Antipolis).[1] With the exception of Sophia-Antipolis, these regions have a high density of universities and research

Table 4.1   Typology of regional influences

Location
- spin-offs
- sources of foreign capital through inward investment

Innovation
- technology transfer/innovation
- information resources
- localise foreign technology
- technological spill-over

Labour
- mix of labour skills
- training

Identity
- contribute to cultural characteristics of the region
- refocusing of region/spatial and technical segmentation or integration
- prestige
- participants in territorially organised policy processes

institutes. Helsinki, for example, has the greatest concentration of universities and other high level expertise in Finland and Munich is similarly dominant in Germany (Sternberg and Tamásy, 1998, p. 7). Grenoble, Oxford and Cambridge have higher densities of knowledge institutes than might be expected from the size of their population. This reflects their historical importance in national innovation systems.

The sets of relationships in which knowledge institutions participate within their local hinterlands are related to the measures they adopt to facilitate co-operative relations with local firms and to encourage spin-off. In some of the regions studied these are the result of centrally planned policies, in others the initiatives of individual institutions play a more important part. It will be seen that most universities have developed radically new interfaces which offer a large range of services ranging from information on core competencies, through the provision of specialist testing and related services, to contracts for joint research. Countries and/or regions which impose specific objectives on institutions or have taken measures to create interactive systems include Spain, The Netherlands and Germany. Those which allow more autonomy to individual institutions comprise the UK, Finland, and Sweden. Case studies from France provide examples of both central planning and local initiative.

4.3.1 Centrally Determined Actions

In Spain, Catalonia has low levels of university and industry R&D for a region at its stage of development partly because many of the decisions are made outside the region. On the other hand, the University Reform Law of 1983 made external links easier and more flexible by removing the legal obstacles to such collaboration and by permitting teaching staff to receive payment for their work on joint private-public projects. This, together with the subsequent creation of Technology Transfer Centres (CCTs) in the region's universities, made company/university links in general and company/university R&D links in particular far more dynamic and market-orientated.

In Barcelona, the OTRI (Office for the Transfer of Research Results) and the university-company liaison offices (CTT) provide science-technology-industry contacts. They are charged with identifying transferable R&D, collaborating in the negotiation of contracts and agreements, administering patents and bringing together R&D needs of companies with the research orientation of university laboratories. OTRIs were established in 1988. Now 77 are operating throughout Spain which together have $180m a year in contracts and provide $13.3m worth of services. Barcelona, with four public and two private universities, is the most important centre, accounting for a quarter of the contracts. However, few local SMEs link into their university for their science and technology competencies; product innovation is more important than process innovation; quality is a major issue; and R&D linkages are based on informal rather than formal contacts. In general, universities and public sector technology transfer offices are evolving from being purely administrative to adopting more aggressive commercial activities and dissemination of the technological capabilities of the universities, moves which if anything have widened the knowledge gap between research institutions and small business. Nevertheless, the efforts by universities to approach companies have probably been greater than the efforts of firms to approach universities (Escorsa, Maspons and Valls, 1998, p. 5).

In France, the development of the Sophia-Antipolis Science Park began in 1971 with the decentralisation of public research institutes from the Paris region. The National Institute for Research in Information Technology and

Automation (INRIA) played a fundamental role in the emergence of an innovative capacity by attracting research competencies and generating start-ups. However, this was not sufficient to support the development of a labour market for highly qualified workers or to lessen the dependence on external resources. In this important respect, the relocation of research institutes and doctoral studies of the University of Nice to Sophia-Antipolis from 1985 was a defining moment. Concurrently, associations and clubs were being created to promote linkages and transfer of information and technology. For example, Telecom Valley, a network headed by INRIA and France Télécom research institutes, brought together firms (large and small) and research centres in telecommunication. This platform continues to promote new technologies and aids technology transfer by opening up large firms to other members and by sharing technological problems and resources. Other initiatives (Eurosud 155, for example) are also important for SMEs.

In Germany, the engagement of universities in the local economy is a function of public policy. In Munich each university and technical college has an agency responsible for technology transfer within the high-technology region of Munich as well as to all other areas of Germany (Sternberg, 1996, p. 43).

The emphasis on public sector R&D in the Netherlands has encouraged a strong interface between public research and the private sector, including SMEs, motivated by the Dutch government's recognition of lack of durability of co-operation between firms and universities. This was underlined by a McKinsey report (1997) which indicated that the failure of universities and new enterprises to co-operate constituted a serious barrier to growth in the Dutch 'software-industry' (Wever, 1998, p. 4).

Finland has developed policies to co-ordinate university research and the commercial activities of firms based on the technology push tradition. The most important is the Finnish Spinno Programme established in 1985. This fosters spin-offs based on researchers' research results and works to commercialise new business ideas by training and consulting.

4.3.2 Individual Institutional Mechanisms

In the UK, since March 1985, universities have taken over the responsibility of exploiting the intellectual property generated by Research Council research from the British Technology Group (BTG). However, industrial liaison activities preceded this by some years, with some having formal institutions years before others. Of our two case studies, Cambridge University established technology transfer institutions far earlier than Oxford University. In 1970 it set up the Industrial Liaison and Technology Transfer Office (now the Wolfson Industrial Liaison Office) to help academics to commercialise their research. It also operates Lynxvale Ltd, the University's technology exploitation company. The University is also involved in Cambridge Research and Innovation Ltd and Quantum Fund, local investment funds for university scientists. Other Cambridge initiatives include Trinity College's Cambridge Science Park (1970), and St John College's Innovation Centre (1987). Generally, Cambridge University has moved from *a climate of disapproval* to being 'positive about entrepreneurial academics' (Cambridge University Reporter July 1990, quoted in Garnsey 1992). In Oxford formal arrangements were not put on a full-time basis until the late 1980s. ISIS Innovation, wholly owned by the University of Oxford, was established in 1988 and a full-time industrial liaison officer was appointed in 1989. ISIS Innovation's main activities are the handling of downstream intellectual property rights issues, dealing with patents and licenses, managing the Oxford Innovation Society, a meeting point for academics and business, and assisting in the formation of new firms.

The significance of these arrangements for the development of local relationships was examined by the comparative study of high-technology SMEs in Oxford and Cambridge carried out by the ESRC Centre for Business Research of Cambridge University in 1996 (Lawson et al., 1998; Keeble et al., 1998). Slightly more Cambridge than Oxford firms had heard of the industrial liaison and or technology transfer services (64 per cent, compared with 58 per cent), but Oxford firms had found them more useful (16 per cent compared to 4 per cent). However, when firms were asked whether they had problems in dealing with universities, rather more Oxford firms (37 per cent) than Cambridge firms (21 per cent) had found difficulties. This suggests that the more formal environment in Oxford was inhibiting the development of a socio-technical milieu such as that found in Cambridge which eased barriers to technology transfer.

In Grenoble, there is a laissez-faire attitude to the industrial exploitation of university-generated science and technology. The history of transferring research results to industry reaches back to the beginning of the 20th century, shaped by the invention of hydroelectricity. It was reinforced after the Second World War, particularly for magnets production, and the tradition continues with a considerable capability for producing new business activity linked to research centres (de Bernardy and Loinger, 1997, p. 254).

4.4 Knowledge Institutions and Cluster Evolution: Interaction and Technology Development Processes

This section examines the contribution made by universities and research institutes to the evolution and composition of high-technology clusters, through processes of interaction and local knowledge development. These include firm spin-offs, attraction of inward investment, innovation stimuli, movement of highly-qualified staff, image creation and the establishment of science parks.

4.4.1 Spin-offs

The establishment of knowledge-based organisations with novel ways of interacting co-operatively and encouraging academic entrepreneurship is a major contributor to local economic development in several of the regions studied. The most striking example of spin-offs is Göteborg, closely followed by Helsinki. In both Göteborg and Helsinki academics are actively encouraged to create new businesses. In Finland, as in Sweden, the intellectual property rights to inventions made in universities reside with the researchers who are then favourably placed to develop their own businesses (Autio, 1998). In both cases, this encourages new firm foundation and close linkages between SMEs and local research institutions.

In the Göteborg region local universities had, up to 1995, spun-out approximately 350 firms, of which around 240 came from Chalmers University of Technology. The links with the University remain close as spin-offs often use university equipment, and may even house their own equipment (especially if bought with grants) at the university (Lindholm Dhalstrand, 1997, p. 50). In 1993, 87 per cent of the start-ups prior to that date were still in business or had been acquired. The direct USO (University Spin-Out) bankruptcies, nearly all during the recession years (1990–93), amount to about 1.4 per cent per year, well below the regional rate of 9.6 per cent. In 1992, 116 'indirect spin-offs', established by former university staff or students after a period of private employment, were documented. Finally when compared to corporate spin-offs, Chalmers USOs grow more slowly but are more successful in terms of patenting activities (Lindholm Dahlstrand, 1998).

The Spinno Programme supported the formation of some 120 new, technology-based enterprises by the end of 1996 and only one had failed. These companies are small but the belief is that the catalysing affect of such new, technology-based companies is even more important to development than the organic growth of existing companies. The main sources of Spinno entrepreneurs have been Helsinki University of Technology (24 per cent), the University of Helsinki together with Helsinki University Central Hospital (22 per cent), the Technical Research Centre of Finland (15 per cent), the University of Industrial Arts Helsinki (12 per cent) and Helsinki School of Economics and Business Administration (6 per cent). Every year, some 30 participants in the training programme are carefully selected from a large number of applicants. The future objective is to help establish 50 new, technology-based companies. In spite of the large number of successful spin-offs, the Programme has been less successful than anticipated. Rather more effective in encouraging growth has been the 'Service Vouchers for SMEs' initiative which offers the work of a graduating engineer to solve technical development problems and which has attracted 60 companies yearly (Kauranen and Makela, 1998; Autio, 1998).

Grenoble too has a tradition of spin-offs. More than 75 spin-offs from research institutes and universities can be identified as being formed since 1982. The period after the second world war had seen significant start-ups directly transferring research results (Sames and Ugimag for magnets). In the 1960s and 1970s, fewer spin-offs appeared: AIM, Mors in 1962, both from INPG. Mors was acquired and developed by Telemecanique. Following acquisition, the company generated 15 spin-offs in a very short period (1972–74). These spin-offs took advantage of opportunities offered by the ZIRST Science Park including cheap rents. These firms were some of the first firms on the ZIRST and were in at the origin of a cluster located on the ZIRST devoted to software and hardware, which became one of the most important specialities of Grenoble. The ZIRST itself came to play a major part in micro-firm formation. Supporting spin-offs was programmed into the CEA-G as the means of transferring research results to production and because of the low level of interest firms had in researchers' 'radical' ideas. An incubator, ASTEC, launched in 1986 has created 40 firms in 10 years mainly around electronic devices. At the INPG High School, the Hitella incubator is devoted to helping existing firms co-operate closely with researchers. This kind of platform has also developed at the CEA-G, for example, testing materials. And GRETH, a club of some 100 high-technology firms, is managed by a team in thermic research and application. Over the years, these initiatives have significantly influenced local developments, particularly as original spin-offs have spawned two or three more to exploit new product ideas (de Bernardy, 1998).

Sophia-Antipolis has also developed spontaneous spin-offs in the 1990s. The first high-technology SMEs came from INRIA (for example Symulog, Ilog Sodeteg) in software, modelisation and image processing. There is now a cluster of high-technology SMEs in this sector (for example Acri, Espri Concept, Géoimage, Isa) with strong inter-firm and research linkages. This is promising for local development because these firms are capable of promoting a local industrial network of service subcontractors (editing, translation, etc.). INRIA, whose mission is to foster firms by providing research facilities, has also created very small leading-edge high technology firms which have been acquired by SMEs creating an alternative and efficient means of technology diffusion. The university has spun out some successful firms including Acces Privilège, and similar activities are observable in health sciences in the university and INSERM (Longhi, 1997).

Both Oxford and Cambridge Universities have adopted a laissez faire approach to spin-offs but with less obvious success than in Finland and elsewhere. The 'Cambridge Phenomenon' results more from indirect spin-offs than from those directly from the university. Thus Cambridge Consultants (now a subsidiary of A.D. Little) was established in 1966 as a spin out from the University and in turn has spun out PA Technology which spawned Scientific Generics, Technology Partnership and Symbionics. These consultancies together employ 1200 highly qualified research and technical staff in the local area and they are the source of a number of new technology-based firms. Both direct and indirect spin-off activity are also important in Oxford where there are university spin-offs with first and second generation off-springs. The origins of some 60 firms can be traced back to Oxford University, of which around 50 survive and employ 4000 people. A feature common to both universities and research institutions in the Oxford and Cambridge regions is a specialisation in bio-medical research and there is a growing population of university spin-offs in this sector (Lawton Smith, 1997a; Keeble and Moore, 1997).

In contrast, some regions have produced very few university spin-offs. In Munich as in Milan, Barcelona and Utrecht, relatively few new SMEs have been generated by local universities. Only about 6 per cent of the high-technology SMEs in the Munich region have been founded by LMU graduates from Ludwig Maximilian University. And in Munich the existing technology transfer facilities are nowhere near fully utilised (Sternberg and Tamásy, 1998, p. 8). Milan also provides the exception to the general tendency in the case study regions for universities to play an important part in local high-technology development. There, the growth of the high-technology SME sector has been largely independent of the university. However, in the Pisa region there is evidence of important links between the university and local high-technology local firms (Capello and Camagni, 1998), which re-inforces the message that clusters of industry-academic links are context dependent.

A comparison of the TSER European network studies yields the interesting finding that in four of the regions where data are available, almost identical proportions of the total regional stock of high-technology SME are university spin-offs. Thus approximately 17 per cent of surveyed firms in each of the Göteborg, Italian (Piacenza, Pisa and North-East Milano), Cambridge and Oxford regions were set up by founders originally employed in local universities. The fact that the definitive early-1980s study of the Cambridge high-technology 'Phenomenon' also found exactly this proportion might possible indicate that this is a steady-state, equilibrium share of any evolving European regional high-technology cluster centred around a major university. As yet we have no explanation why a lower share was recorded by the Munich region survey (Sternberg and Tamásy, 1998) or why there were no university spin-offs in the Utrecht sample (Wever, 1998).

4.4.2 Inward Investment

Perhaps even more important in terms of employment growth, technological investment and cultural identity are the extent and kind of inward investment by multi-national companies in high-technology clusters. Inward investment has been and is increasingly important in Oxford and Cambridge, and was the *raison d'etre* for Sophia-Antipolis. In Oxford and Cambridge inward investment has tended to take the form of research facilities established by major US, Japanese and other European corporations. In Cambridge, research laboratories are operated by Schlumberger, SmithKline Beecham, Toshiba, Sony and Microsoft. Oxford University has a similar record of attracting firms into the county, including foreign firms such as Yamanouchi (pharmaceuticals) and Sharp (electronics), and UK firms such as British Biotech and SmithKline Beecham (biotechnology/pharmaceuticals). In Sophia-Antipolis, the idea was to attract capacity into the park. There has been considerable success. High-technology SMEs have moved in to sustain their growth and/or to locate near the sources of their technologies. E3X, for example, is developing Unix technologies in the domain of Telecoms, and has moved in to find space for growth, and because its technologies came from INRIA.

The R&D centre of E3X grew from 6 to 48 persons between 1991 and 1994. The firm has been acquired by Télésystèmes (a subsidiary of France Telecom) which has adopted new technologies by means of the acquisition. Other high-technology firms have moved to Sophia-Antipolis for a pure image effect and they are less likely to network (Longhi, 1998).

4.4.3 Innovation

In the definition of an innovative milieu, proximity contributes to synergistic effects and in the case of the relationships between knowledge institutions and local firms this would take the form of contributing to their innovative capacity. However, the evidence is not clear cut. Some knowledge institutions are playing a more active role, while others are only at the stage where linkages are becoming important. Moreover in some places local interaction tends to be a large-firm phenomenon, for example in the Netherlands, but in others, for example in Finland, SMEs have been able to overcome traditional barriers associated with interacting with universities. In spite of the evidence of the overall importance of linkages, the main finding is that firms in each of the regions surveyed by network members claimed that their customers were more important external sources of innovation than knowledge institutions. This raises important questions about the balance between local embeddedness of technology transfer and links with partners outside the locality in the evolution of high-technology clusters and how this in turn affects the structure and quality of links within the locality.

Grenoble is an example of a region where government laboratories and universities have traditionally played an important role in stimulating technical development. Since the beginning of the century, relations between knowledge centres and industry have been of central importance in stimulating, and renewing, innovative activity. For example, in telecoms, the CNET (Telecom public research centre) was of overwhelming importance for the development of SMEs in Grenoble's technology park. The volume of agreements between knowledge institutes and firms (600 in 1996) is important but it is also important to note that fewer than 25 per cent of local firms have such links.

Munich provides a second important example of well-developed relations between business and higher education institutes. There, parallel evolutions of dual training and technology transfer have been mutually reinforcing. Links are particularly strong and well organised around the

Ludwig Maximilian University (LMU) and other governmental laboratories including the Fraunhofer Society. Technology transfer is mainly within the region: i.e between knowledge centres and firms located in and around Munich. A recent study of 300 small high-technology firms in the Munich region shows that almost one-third of the SMEs maintain contact with LMU; limited project work is the most common form of co-operation; half of the co-operating businesses work together with the LMU in the R&D field; the great majority of the businesses are pleased with the joint ventures; and lack of information about the research potential of the LMU is the most important reason for the absence of co-operative links (Sternberg and Tamásy, 1998).

The findings from the Sophia-Antipolis study provide support for evolutionary theories which specify the interdependence among actors (see Storper, 1997). Here the evolving role of knowledge centres in the development of a local economy is related to a combination of state, institution and firm activities. The bringing together of firms and institutions in Sophia-Antipolis in IT, telecoms and electronics resulted from public planning. Originally, however, local interactions were sparse and if anything the firms and institutions relocated to Sophia-Antipolis themselves drew in and developed local high-technology capability rather than capitalising on what was already there. The turnaround came when the university effectively entered into the park. Within a few years this had a cumulative effect as more and more SMEs were created. The role of the university is also pronounced in the local labour market for qualified workers with the evolution of cultural proximity and personal networks of graduates employed in the park's firms. With these developments Sophia-Antipolis has become increasingly integrated with its locality further justifying the large expense of decentralising the public research laboratories and 'great schools'.

In the Netherlands there has been *quasi-integration*. Wever's study (Wever, 1998) of 94 SMEs in the Utrecht and Rotterdam regions found that only 21 per cent (20 firms) considered knowledge centres as important for their innovation activities. These links are both informal and formal although the latter dominate. Most firms with links with knowledge centres anywhere also had contacts with those in the region. It seems easier for larger (10–100 employees) and more innovative firms to contact knowledge centres than smaller and less innovative firms. Only 20 per cent of the smaller IT-firms (less than 10 persons) had links with knowledge centres, against 46 per cent of the larger IT-firms.

In Finland, Autio (1997) found that two thirds of the high-technology SMEs on the Innopoli-Otaniemi Science Park surveyed had close links with Helsinki University of Technology and more than half had links with the Technical Research Centre of Finland (VTT), a national technology transfer organisation. SMEs were also found to be active in using external technology services and financial support; around one third had used the VTT and Helsinki University of Technology (HUT). An important finding of the Autio study was that services tailored to the needs of individual firms are in higher demand than general 'infrastructural services'. The organisation of the latter is based on the traditional *technology push* thesis and does not efficiently support the type of targeted services needed. These require more direct relations between SMEs and the suppliers of new technology.

4.4.4 Labour Mobility

The impact of universities on the quality of milieux through the development of localised markets for highly skilled labour is country and/or institution specific. Where universities are long established they serve a cultural and technological purpose in their education and career promotion roles by which links are forged with firms. This is important in all regions but particularly so in Munich, where the dual training system provides an interactive link between higher education and industry, and in Grenoble, where the university has played an important part in the 'renewal of competencies' by involving the staff of local high-technology firms in their training courses. By contrast, Oxford University has not by design played a role in local economic development. It has not consciously created career paths for graduates or employees into local high-technology firms through either the spin-off process or the matching of undergraduate and graduate training to the needs of local specialisations such as biotechnology. On the other hand, Oxford Brookes University in 1997 established a motorsport engineering degree supported by the local motor sport community with the express purpose of overcoming local skill shortages in that industry (Lawton Smith, 1998).

One indicator of the availability of qualified employees in the vicinity is the extent to which firms recruit locally and have staff with degrees from local universities. Compared with Oxford, the Cambridge firms surveyed in the ESRC Centre for Business Research study noted earlier had better qualified managers and directors but in Oxford research staff had superior qualifications. In spite of the absence of formal systems designed to enhance local recruitment, more Oxford than Cambridge firms recruit from the local university; but in Cambridge inter-firm mobility is more important, particularly for research staff (Keeble et al., 1999). In Oxfordshire, government laboratories were an important source of skilled craftsmen for high-technology firms.

Local recruitment is highly developed in both the French case studies and in the Swedish case. In Grenoble 25 per cent of new engineers (900) graduating from INPG found their first job locally, encouraged by the participation of firms' staff in courses run by the university. In Sophia-Antipolis, the contribution of universities to the scientific labour market has been of increasing significance since the re-location there of research institutes and doctoral programmes of the University of Nice. This increased training capability and reduced reliance on an inflow of highly qualified workers facilitated the emergence of a local labour market for highly qualified workers and encouraged the development of local industry-research linkages in which students play a significant role (de Bernardy, 1994; Longhi, 1997). Similarly, in Göteborg the important role of Chalmers University as a provider of graduates adds to the large number of spin-out firms as the driving force in the development of the local labour market.

4.4.5 Image

Manifestations of the cultural identity and the image of regions accumulate through a series of formal and informal processes. In some regions the nexus of high-technology firms and knowledge institutions dominate the character of the region. Notable examples are Sophia-Antipolis, Cambridge and Grenoble.

In Sophia-Antipolis informal links are perhaps more important than formal relationships in creating the atmosphere for and the possibility of information exchange and co-operation. Cultural activities around universities (for example in Cambridge) add to cohesion. These are far less in evidence in Oxford. In Grenoble, moves such as the Grenoble Network Initiative encourage involvement in the shaping of the future of the city. Networks derived from research activities and recruitment bring together a variety of skills to enhance problem-solving capabilities. Furthermore, the participation of employees of Grenoble firms in university education helps develop co-operation in research. In these ways, knowledge centres crystallise and foster local professional identity and innovation. There is also evidence that universities have an important role in reconstituting local labour markets particularly in areas where large international companies are located. Labour market participation facilitates a sense of common identity.

4.4.6 Science Parks

Science parks are very much part of the projected image of regions and in some places contribute to the process of interaction between industry and knowledge institutions. They have two broad functions: 'land use profitability' and 'commercialisation of the science base' (Moore, 1997). The former is of greater importance in both the Oxford and Cambridge science parks. However, the St John's Innovation Park in Cambridge seeks to perform both functions, although by linking firms to other firms rather than to the university. The Science Park in Oxford exploits its links with the University, rather than vice-versa, for example by emphasising the connections in discussions with potential inward investors. The Oxford Science Park was established by Magdalen College, two decades after Trinity College founded the Cambridge Science Park in 1970. This delay in Oxford hindered the development of a high profile image such as that which developed in Cambridge in the 1980s.

In Grenoble, the idea of setting up a technology park as a means of encouraging the commercialisation of the results of local research was proposed in 1969. The park, the ZIRST, was established in 1972 and became fully operational in 1974. It now accommodates 220 high-technology firms, more than 65 per cent of which are SMEs. ZIRST is not directly managed by the research institutes which, nevertheless, actively participate in decisions on new entrants using selection criteria based on the innovative potential of would-be tenants. Other parks in Grenoble have prioritised the 'commercialisation' of the science base and even in older industrial parks the population of high-technology SMEs is increasing, particularly around the university campus. The ZIRST and Sophia-Antipolis science parks are contemporary and in both the importance of public research funding is a mark of state leadership. This was originally driven by science push but this orientation has had to be reconsidered with the trend to more customer pull. In the Göteborg and Bohus region there are a number of technology centres and new ones are being developed; two laboratories for biochemistry have been added. There are both science parks and innovation centres in the region and more are being established. In Uddevalla, for example, there is a recently established Business and Innovation Centre (BIC) with partial funding from the European Commission.

The Helsinki Science Park has a specific sectoral profile. It focuses on biotechnology with an incubation unit and incorporates Biomedicum, a new biomedical science park. Innopoli Otaniemi Science Park has extended the interface between Otaniemi research institutions and Finnish industry. Tenants of this science park act as catalysts by utilising the research institutions' technical findings, developing them and injecting them into the products and processes of their customers. High-technology firms surveyed by Autio (1997) regard tapping into new technology as the most important science park benefit; the use of university students as a part-time workforce is the second most important. The science parks tend to attract small-growth high-technology firms (50 per cent being direct university spin-offs) and the occupancy rate of incubator sections remains over 80 per cent despite strict selection criteria. However, Autio suggests that there is an over supply of general information and assistance but a paucity of the hands-on, targeted support most small firms need, perceptions of the success of which, Autio's study suggests, may be optimistic. In general although science parks have been established in Finland to emulate the experience of Silicon Valley, the reality is that science parks are often set up as responses by knowledge centres and land planning services to plant closures and redundancies. The way forward may not be so much to attempt to solve the problem of de-industrialisation by reproducing success stories but to foster more effective ways of technology and knowledge transfers incorporating both science push and market pull.

In Munich much less use has been made of science parks as a focus for high-technology industrial development. It has no university science park but it does have a government-sponsored technology centre (Münchner Technologie Zentrum) which houses 60 more or less RTD-intensive SMEs (Sternberg and Tamásy, 1998). These authors question whether science parks provide special advantages which are not available elsewhere.

4.5 Oxford and Grenoble: Similar but Different

A comparison between Oxford and Grenoble is instructive in highlighting the different ways by which areas develop. Their similarities as the locations of their countries' nuclear research laboratories and increasingly of leading edge research in their high technology firms disguises structural differences. Whereas Oxfordshire's economy has undergone a considerable transformation since the 1960s, diversifying from an agricultural region which also had a car assembly industry, to a high profile centre of research in the public and private sector, Grenoble has been a centre of large scale energy, production and research activities since the start of the twentieth century. The establishment of new and relocation of existing major government research facilities in Grenoble has taken place throughout the century while the atomic energy laboratory was only established in Oxfordshire in the 1940s and 1950s.

Grenoble is second only to Paris as a research pole, particularly in science for engineering. The primary role of research is notable in the Grenoble case, where the impact of research laboratories such as LETI or CNET (electronic engineering) has been significant in the development of the region's complex of micro-electronics firms. Similarly, in Oxford, the university's influence underpins the existence of the region's recent biotechnology/life sciences specialisations and was the origin of Oxford Instruments which spawned the local cryogenics industry. There is, however, an important difference between Grenoble and Oxford concerning the role of atomic energy and other national laboratories. In Grenoble these have contributed to the technological development of the city, whereas in the Oxford region the national laboratories and Oxford University have remained more remote from local industrial development until the 1980s at the earliest.

Grenoble has four universities, in which INPG (8 engineering schools) has 48000 students of whom 5500 are foreign. Five public research centres (CEA-G, CNET, CNRS, CRSSA, INRIA) and five European research centres (ESRF, ILL, IRAM, LCMI, EMBL) employ almost 10000 public researchers in 220 different laboratories with a broad span of activities. But, most importantly, knowledge institutes are, and have been for almost a century, actively involved in the local economy. The Atomic Energy Authority (CEA) laboratory LETI was established in 1967 with the express purpose of commercialising research undertaken in the nuclear programme. It specialises in microelectronics and information technology. LETI did not hesitate to develop new practices without the explicit agreement of CEA's management in Paris; a degree of autonomy which promoted pioneering work in physics and derivative activities. However, attempts to force the pace of change, through, for example, programmes to encourage spin-out in LETI, have not had a dramatic effect. Rather there has been a steady flow of spin-offs serving as one of the many sources of transformation for the local economy. The majority of researchers seem to remain committed to research for the creation of knowledge with potential for application rather than looking to leave research to exploit particular technical opportunities by the spin-off process. Nevertheless, LETI appears to be one of the most efficient of the laboratories in the creation of links with firms and has a clearly assigned role in the fostering of innovation locally.

These and other developments have led to an increase in the match over time of the technical capacities, culture and pace of production between the research institutions and local firms. One of the first new firm incubators was created in CEA-G in 1986 and many agreements had been concluded between firms and universities. Specific associations had been established to develop these links. ADR (Association for the Development of Research) was set up in 1929 to act as manager and employer for the development of collaboration and fostered the growth of research capacity by private contracts. AUG (University Alliance of Grenoble), created in 1947 by Paul Louis Merlin, the head of a leading local firm, is devoted to promoting links with firms. AUG has taken the initiative in creating accommodation for students, soliciting grants from firms for students, funding experimentation and creating committees for considering matters of common interest. It acts as a locus of collective learning by offering opportunities for the exchange of knowledge between individuals and also fosters collective experimentation. The ability to establish common practices by interaction is still in progress as a tacit method to experiment with new programmes. This is demonstrated by the development in 1980 of SILIU, an industrial Liaison Office at the scientific university. SILIU's role is to seek the best laboratory to meet the needs of the firms. It can also manage contracts and provide the means of contacting individual academics. However, researchers and industrialists often by-pass these contact points and prefer more informal ways of meeting: reputation, student projects, conferences and fairs, etc. This *atmosphere* is of prime importance in understanding why Grenoble has chosen an economic strategy more orientated towards high value and customised intermediate products than to mass production. This scientific profile has grown through effective co-operation between knowledge institutions and firms which remain connected to the university community. Moreover, despite the limited evidence of formal links, relations between firms and knowledge institutions are very important and these provide ways of confronting problems and ensuring a renewal of activities in the face of set-backs.

Oxfordshire's science and technology knowledge base comprises public investment in the form of four universities, nine hospitals of which seven have University research departments, and some 10 nationally important research laboratories. Oxford University is the second largest after London with a combined undergraduate and graduate population of nearly 15,000. Oxford has remained more research focused than Grenoble but that is changing. The key factor in shaping the outlook of the national laboratories towards technology transfer was the regulatory change which first gave agency status to the United Kingdom Atomic Energy Authority (UKAEA) laboratories and then privatisation of the commercial arm AEA technology in 1996 (Lawton Smith, 1997b). Oxford University is also undergoing rapid internal change. In mid-1997 a new head of ISIS Innovation was appointed and a new appointment was made within the Research Support Office in 1999. These changes, particularly those to ISIS, are likely to have a considerable impact on the way in which intellectual property is exploited. An Oxford University Technology Transfer Committee's Report published in 1995 was highly critical of the current lack of co-ordination of industrial liaison activities and support for would-be entrepreneurs. The Review also acknowledged that some academics had formed their own companies independently of the university system. As a response to the report, Oxford University has enacted a Statute of claiming ownership of IPR generated by its staff and students in the course of or incidental to their studies. This replaced the previous Statute under which the University 'asserts its ownership of them'. The new Statute became effective in July 1996 (University of Oxford, 1996). The changes within ISIS and the structures resulting from the Report's recommendations will take time to have an effect on the university's relationship with the local high-technology economy. The implications of the Statute are that individuals will have less control over the exploitation of the results of their research if the capitalisation of IPR becomes more bureaucratised. This may discourage some potential entrepreneurs. On the other hand, if the new systems are properly resourced (which they are not yet) and information diffusion is encouraged, perception within the University of the routes to commercialisation may be heightened and closer links between the university and the local high-technology economy encouraged.

The cases of Grenoble and Oxford are important examples of ways by which knowledge institutions have been obliged and are attempting to adapt to the change in their environments. Grenoble has evolved more institutionalised systems designed to integrate its national laboratories and universities into the local milieu than in Oxfordshire. As such they provide indicators of the opportunities and limitations within which they have to operate. It is interesting to note that decades ago in Grenoble a minority of actors within universities and research centres began to develop closer links with firms, paid greater attention to industries' technological needs, placed greater emphasis on the production of applied knowledge, developed professional and vocational education as a way of maintaining employment, and developed alternative funding to that provided by the state; priorities which virtually all knowledge institutions now embrace.

4.6 Evaluation and Conclusions

The network's comparative study of European clusters of high-technology SMEs reveals a wide range of different processes by which knowledge centres (universities and publicly funded research organisations) are continuing to be or becoming important as regional collective agents. The evolution of the labour market is especially significant in the incorporation of firms and organisations into collective learning processes. The comparative case studies also lend support to the argument that in many, if not most, of Europe's technology-based innovative milieux there is some mapping of technology specialisation, sectoral emphasis, and strengths of local knowledge centre on to adjacent SME clusters.

The impact and effectiveness of technological spill-over varies and is changing. However, generally knowledge centres, as providers of new technology, participants in learning processes and networks by which technology is transferred, and as the trainers of scientists, technologists and engineers, play an important part in shaping and reshaping economies. The evolution of some areas is particularly impressive when judged by the growing contact between university-level research and industry and by the creation of intefaces which increase the number and widen the range of firms associated with research institutions. In other cases, there is a greater degree of independence between the development of SMEs and high level research activities. These various types of interaction change the forms of knowledge held in the locality as technology is translated from the academic to the commercial domain and where feed back loops influence the direction of research. The range of means by which these interactions take place contributes to the formation of differing characteristics and components of local knowledge-clusters so it becomes plausible that endogenous factors become more important over time than exogenous ones (Lorenz, 1997, p. 14).

Case by case the evaluation shows that:

- research and education institutions in Sophia-Antipolis and Grenoble came to play major integration roles at the regional level and in managing relational networks for the inclusion of SMEs in innovative projects. The difference between the two is that whilst the iteration between scientific research and industrial development is intrinsic to the historical trajectory of economic development in Grenoble, in Sophia-Antipolis it was triggered by the insertion of institutions of higher education into an agglomeration of large public research laboratories and R&D departments of large firms.

The absence of significant research competencies in biotechnology within Chalmers University of Technology arguably explains the relative absence of SMEs in this sector in the Göteborg region. However, there is also evidence that over time and with growth and increasing maturity of a regional technology cluster, the sectoral evolution and developing technological expertise of a region may become increasingly detached from the historic strengths of the local knowledge centres. Thus in the Cambridge case, while the strong 1970s/1980s growth of computing SMEs (hardware and software) was related to the central role of the university Computer Laboratory and government-funded CAD Centre, the recent mushrooming of telecommunications SMEs in the region (Symbionics, Analysys) as a new and distinctive 'micro-cluster' appears to owe little or nothing to the role of the university, with the new specialisation emerging out of local firms and R&D units. On the other hand, recent biotechnology growth, another new regional specialisation, does to some extent reflect university and research institute influences. Informal links, labour skills, spin-offs and credibility, prestige and reputation by association, collectively appear to represent an important competitive advantage arising from proximity to Cambridge University for technology-based SMEs in the Cambridge region (Keeble and Moore, 1997, p. 16). The importance of university links appears to vary between sectors (e.g., computers, communication technologies and biotechnology) in which the university seems to play different roles.

In Germany the influence of public institutes on innovation in SMEs is more effective because of the pre-existence of local linkages forged by public and private co-operation. Munich is notable for the development of formal relations managed through 'ad hoc' institutions whilst the spin-out of new firms from the region's universities remains low.

In all areas, research in universities remains to a large degree separated from local industrial development except in relation to larger firms which have their own research capability enabling the more ready transmission of new knowledge. Knowledge centres are in no way accountable to SMEs who have no ways of influencing the direction of research. The lack of such capacity helps explain both the limited technological interaction SMEs have with knowledge centres and their need for more *overt* co-operation. The high rate of university spin-off in the Göteborg case and its official encouragement means that the potential for co-operation is high because of the firmness of the continuing links between the new firms and the institutional origins of their founders. The effect is a new generation of research-intensive firms willing and able to develop high levels of interaction with academic researchers which is creating a culture of co-operation with the universities. Such a common culture appears to be evolving in Helsinki although firms there claim that the origin of their ideas is elsewhere. However, the Finnish research highlights the need for a change in emphasis of policy away from the general and top-down encouragement of technology diffusion to a more targeted attention to the specific needs of small firms. These may be analogous to the internal requirements of large organisations as they move from the initial conception of new products and processes to proto-typing and full production.

Figure 4.1 illustrates the demand on both sides for an interface between education and knowledge centre research and local business. The contention is that the linear, sequential, technology push view of innovation obscures the nature of industrial demand for the technical services of knowledge centres. As the locus of high-technology development shifts toward network structures, knowledge institutions are responding to new challenges that insert them into the market as economic agents.

This suggests a need for more autonomy of individual laboratories and their researchers. In practice, this would mean diverting an increasing proportion of the financing to the contact (supply and demand) interface depicted in Figure 4.1. Achieving this would require greater freedom for knowledge centres to manage their resources; a strategy adopted to varying extents in the regions studied by the network. The move towards creating high-technology SMEs more directly involved in research and better informed about research programmes of knowledge centres is of great importance in overcoming negative attitudes of both researchers and entrepreneurs to the others' activities and in building bridges between the two. This is of particular importance because we are entering a period when small firms require a greater complementarity between externally-generated research and their business needs, and when technological and knowledge

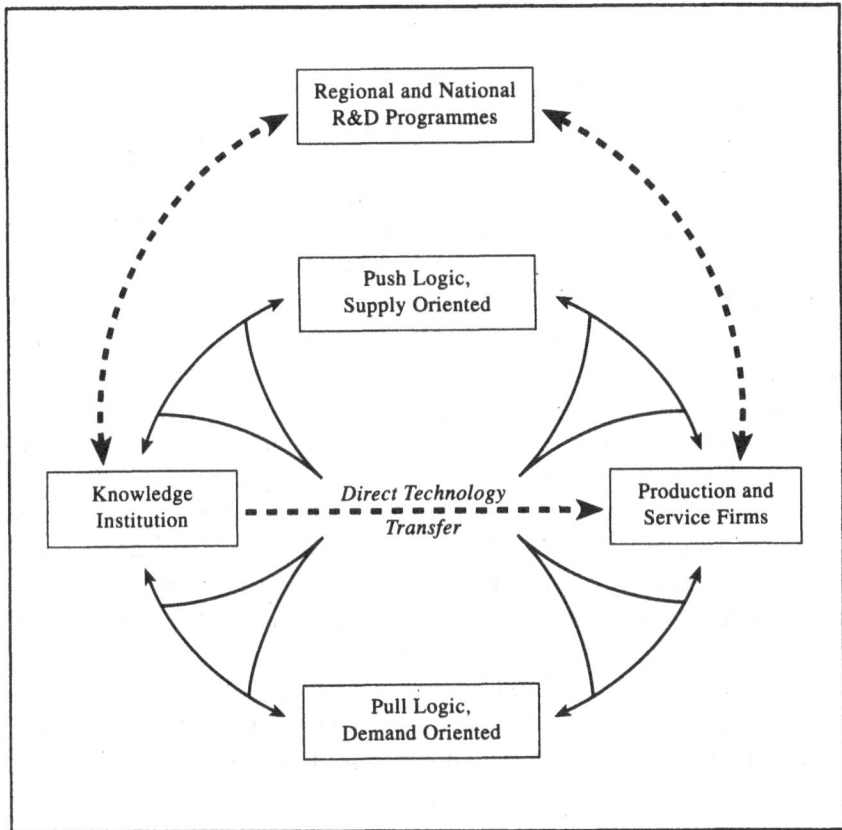

Figure 4.1 Relations between firms and knowledge institutions: a new paradigm is supplanting the old pipeline one

centres need commercial outlets for their research activities to increase their effective use and to make more resources available to them.

It is widely known that for most firms innovation starts from close relations with customers and suppliers. For a third, innovation is triggered by links with other professionals. Three other sources of around 10 per cent include internal research and imagination, observing developments in other firms, and relations with universities and research centres. In highly competitive economies, reliance on customers and suppliers must constitute an insecure basis for innovation. The objective therefore should be to increase the input of innovation from other sources. In this respect, knowledge centres should be indispensable. However, they are a poorly used source of innovation. In overcoming this knowledge institutions have an important part to play in adapting the knowledge they create and improving its diffusion to the mutual advantage of business and knowledge centres.

Developing such a programme requires steering agencies sensitive to both the present and future needs of industry. It is not generally possible for SMEs to manage their short term innovation and long term technology trajectories. Moreover, the innovative process normally requires a mix of competencies, the articulation of which needs to be mutually comprehensible if they are to be effectively integrated. The essential requirements of this for SMEs are spatial and institutional proximity. This is secured in some of the clusters studied by entrepreneurs retaining one foot in the parent knowledge centres and in others by dual training and other long established linkages between research institutions and the business community. These measures serve to more fully integrate SMEs with local innovative systems and to reduce the mistrust and suspicion which helps to explain why in some areas firms innovate independently of universities and other research institutions. The need is for a more cordial atmosphere to cultivate synergy and develop symbiotic relations and in which the independence of each side is also fully respected. In these circumstances knowledge centres can become economic actors by participating in collective learning processes at the regional level while conserving their roles of cultural diffusion and independent, long term, technical and scientific investigation.

Note

1   Tamásy and Sternberg (chapter 3) give the population and size of the knowledge centres in the case study regions.

References

Amin, A. and Robins, K. (1990), 'The Re-emergence of Regional Economies? The Mythical Geography of Flexible Accumulation', *Environment and Planning D*, vol. 8, pp. 7–34.
Anselin, L., Varga, A. and Acs, Z. (1997), 'Entrepreneurship, Geographic Spillovers and University Research: A Spatial Economic Approach', paper presented at ESRC Centre for Business Research Workshop, Cambridge, UK, March 1997.
Autio, E. (1997), 'University Links and Technology-based SMEs in the Helsinki Region', in D. Keeble and C. Lawson (eds), *University Research Links and Spin-offs in the Evolution*

of Regional Clusters of High-Technology SMEs in Europe, ESRC Centre for Business Research, University of Cambridge, Cambridge, pp. 51–82.

Camagni, R. (1991), 'Local 'Milieu', Uncertainty and Innovation Networks: Towards a New Dynamic Theory of Economic Space', in R. Camagni (ed.), *Innovation Networks: Spatial Perspectives*, Belhaven Press, London, pp. 121–43.

Camagni, R. (1993), 'Inter-firm Industrial Networks: the Costs and Benefits of Co-operative Behaviour', *Journal of Industry Studies*, vol. 1, pp. 1–15.

Capello, R. and Camagni, R. (1998), 'Regional Report: Italy', in D. Keeble and C. Lawson (eds), *Regional Reports*, July 1998, ESRC Centre for Business Research, University of Cambridge, Cambridge.

Cooper, A.C. (1971), 'Spin-offs and Technical Entrepreneurship', *IEEE Transactions on Engineering Management*, EM-18 (1), pp. 2–6.

de Bernardy, M. (1994), 'De la High-Tech aux Autres Techniques, Moderniser les Entreprises avec les Laboratoires de Recherche, les Centres Techniques et la Formation', in Territoire d'Europe, Programme communautaire, RECITE, Vol. IV, Junta de Castilla y Leon, FEDER, pp. 143–62.

de Bernardy, M. (1998), 'RTD SMEs and Collective Learning: Historicity and Ability for a Local Economy to Evolve: The Grenoble Case Study', in D. Keeble and C. Lawson (eds), *Regional Reports*, July 1998, ESRC Centre for Business Research, University of Cambridge, Cambridge.

de Bernardy M. and Loinger G. (1997), 'Technopolis and Serendipity. The Adjustment Effects Created by Nancy-Brabois and ZIRST Technological Parks: A Comparison', in R. Ratti, A. Bramanti and R. Gordon (eds), *The Dynamics of Innovative Regions, The GREMI Approach*, Ashgate, Aldershot, pp. 237–63.

EC (1994), *The European Report on Science and Technology Indicators*, EUR 15897, Luxembourg.

Escorsa, P., Maspons, R. and Valls, J. (1998), 'Regional Report: Barcelona', in D. Keeble and C. Lawson (eds), *Regional Reports*, July 1998, ESRC Centre for Business Research, University of Cambridge, Cambridge.

Garnsey, E. (1992), 'An Early Academic Enterprise: A Study of Technology Transfer', *Business History*, vol. 34, pp. 79–98.

Jones-Evans, D. (ed.) (1998), *Universities, Technology Transfer and Spin-off Activities : Academic Entrepreneurship on the Periphery of Europe*, Final TSER report, PL95, 1042, EC.

Kauranen, I. and Makela, J. (1998), 'The Final Report from Finland', in D. Keeble and C. Lawson (eds), *Regional Reports*, July 1998, ESRC Centre for Business Research, University of Cambridge, Cambridge.

Keeble, D. and Lawson, C. (eds) (1997), *University Research Links and Spin-Offs in the Evolution of Regional Clusters of High-Technology SMEs in Europe*, ESRC Centre for Business Research, University of Cambridge, Cambridge.

Keeble, D., Lawson C. Moore, B. and Wilkinson, F. (1999), 'Collective Learning Processes, Networking and 'Institutional Thickness' in the Cambridge Region, *Regional Studies*, vol. 33, pp. 319–32.

Keeble, D., Lawson, C., Lawton Smith, H., Moore, B. and Wilkinson, F. (1997), 'Internationalisation Processes, Networking and Local Embeddedness in Technology-Intensive Small Firms', *Small Business Economics*, vol. 11, pp. 327–42.

Keeble, D., Lawson, C., Lawton Smith, H., Moore, B. and Wilkinson, F. (1998), 'Collective Learning Processes and Inter-Firm Networking in Innovative High-Technology Regions', *ESRC Centre for Business Research Working Paper*, No. 86, University of Cambridge.

Keeble, D. and Moore, B. (1997), 'University and Technology Intensive SME Research Collaboration, Spin-offs and Recruitment in the Cambridge region', in D. Keeble and C. Lawson (eds), *University Research Links and Spin-Offs in the Evolution of Regional Clusters of High-Technology SMEs in Europe*, ESRC Centre for Business Research, University of Cambridge, Cambridge, pp. 15-23.

Lawson, C., Moore, B., Keeble, D. Lawton Smith, H. and Wilkinson, F. (1998), 'Inter-firm Links between Regionally Clustered High-Technology SMEs: A Comparison of Cambridge and Oxford Innovation Networks', in W. During and R. Oakey (eds), *New Technology-Based Firms in the 1990s, Volume IV*, Paul Chapman, London, pp. 181-96.

Lawton Smith, H. (1997a), 'University and Public Sector Research Laboratory Links and Technology Transfer in the Oxford Region', in D. Keeble and C. Lawson (eds), *University Research Links and Spin-Offs in the Evolution of Regional Clusters of High-Technology SMEs in Europe*, ESRC Centre for Business Research, University of Cambridge, Cambridge, pp. 24-38.

Lawton Smith, H. (1997b), 'Regulatory Change and Skill Transfer: The Case of National Laboratories in the UK, France and Belgium', *Regional Studies*, vol. 31, pp. 41-54.

Lawton Smith, H. (1998), 'Regional Report: Oxfordshire', in D. Keeble and C. Lawson (eds), *Regional Reports*, July 1998, ESRC Centre for Business Research, University of Cambridge, Cambridge.

Lindholm Dahlstrand, A. (1997), 'Entrepreneurial Spinoff Enterprises in Göteborg, Sweden', in D. Keeble and C. Lawson (eds), *University Research Links and Spin-Offs in the Evolution of Regional Clusters of High-Technology SMEs in Europe*, ESRC Centre for Business Research, University of Cambridge, Cambridge, pp. 39-50.

Lindholm Dahlstrand, A. (1998), 'The Development of Technology-based SMEs in the Goteborg Region', in D. Keeble and C. Lawson (eds), *Regional Reports*, July 1998, ESRC Centre for Business Research, University of Cambridge, Cambridge.

Longhi, C. (1997), 'Sophia-Antipolis', in D. Keeble and C. Lawson (eds), *University Research Links and Spin-Offs in the Evolution of Regional Clusters of High-Technology SMEs in Europe*, ESRC Centre for Business Research, University of Cambridge, Cambridge, pp. 83-7.

Longhi, C. (1998), 'Processus d'apprentissage collectif à Sophia-Antipolis: l'émergence d'un milieu innovateur', in D. Keeble and C. Lawson (eds), *Collective Learning Processes and Knowledge Development in the Evolution of Regional Clusters of High-Technology SMEs in Europe*, ESRC Centre for Business Research, University of Cambridge, Cambridge, pp. 39-52.

Lorenz, E. (1997), Discussion on 'University Firm Linkages and Technology Transfer: The Minneapolis Case' in D. Keeble and C. Lawson (eds), *University Research Links and Spin-Offs in the Evolution of Regional Clusters of High-Technology SMEs in Europe*, ESRC Centre for Business Research, University of Cambridge, Cambridge, p. 14.

Luger, M.I. and Goldstein, H.A (1991), 'Universities, the Urban Milieu, and Technology Development', Paper presented at 1991 ACSP-AESOP Conference, Oxford.

McKinsey and Company (1997), *Boosting Dutch Economic Performance*.

Moore, B. (1996), 'Sources of Innovation, Technology Transfer and Diffusion', in A. Cosh and A. Hughes (eds), *The Changing State of British Enterprise: Growth, Innovation and*

Competitive Advantage in Small and Medium Sized Firms 1986–5, ESRC Centre for Business Research, University of Cambridge, Cambridge.
Moore, B. (1997), quoted in D. Keeble and C. Lawson, (eds), *University Research Links and Spin-Offs in the Evolution of Regional Clusters of High-Technology SMEs in Europe*, ESRC Centre for Business Research, University of Cambridge, Cambridge.
Sternberg, R. (1996), 'Regional Institutional and Policy Frameworks in the Munich Region', in D. Keeble and C. Lawson (eds), *Regional Institutional and Policy Frameworks for High-Technology SMEs in Europe*, ESRC Centre for Business Research, University of Cambridge, Cambridge, pp. 35–45.
Sternberg, R. and Tamásy, C. (1998), 'Regional Report: Munich', in D. Keeble and C. Lawson (eds), *Regional Reports*, July 1998, ESRC Centre for Business Research, University of Cambridge, Cambridge.
Storper, M. (1997), *The Regional World*, Guilford, New York.
University of Oxford (1996), Intellectual Property Policy Statement of Enactment of Changes to be made to the Legislation Governing Intellectual Property Generated within the University, *Oxford University Gazette*, Thursday 31 October 1996.
Wever, E. (1997), quoted in 'Discussion on University Firm Linkages and Technology Transfer: The Minneapolis Case' in D. Keeble and C. Lawson (eds), *University Research Links and Spin-Offs in the Evolution of Regional Clusters of High-Technology SMEs in Europe*, ESRC Centre for Business Research, University of Cambridge, Cambridge, p. 14.
Wever, E. (1998), 'The Dutch Case: The Randstad', in D. Keeble and C. Lawson (eds), *Regional Reports*, July 1998, ESRC Centre for Business Research, University of Cambridge, Cambridge.

5 The Role of Inter-SME Networking and Links in Innovative High-Technology Milieux

ROBERTO CAMAGNI AND ROBERTA CAPELLO

5.1 Introduction[1]

In the vast literature on territorial clusters of SMEs a strong emphasis is given to inter-firm linkages. This is true for both the well-known theory of *industrial districts*,[2] based on traditional local externalities and on static economies, as well as for the more recent dynamic approach of the '*milieux innovateurs*'.[3]

This chapter provides an insight into the nature of inter-SME linkages in high-technology milieux, and in particular the role these relationships play in the innovative activity and productivity of local firms. The chapter is based both on theoretical propositions derived from the literature on SME clustering and innovative milieux, and on empirical evidence obtained by the research and analyses carried out by members of the European Network on 'Networks, collective learning and RTD in regionally-clustered high-technology small and medium sized enterprises' described in chapter 1.

Section 5.2 of the chapter presents and discusses the theoretical framework for analysing SME linkages. The research issues emerging from the theory are then empirically tested in sections 5.3 and 5.4. Some concluding remarks and further research directions are presented in the final section.

5.2 Inter-SME Linkages in the Theory of Innovative Milieux

Innovative milieux are distinguished from simple agglomeration of economic activities by close inter-firm relations. The existence and importance of inter-SME linkages have been stressed since the early studies of spatial clustering of SMEs based on the 'industrial district model'. This has four basic elements (Rabellotti, 1997), namely:

- a *spatial element*, characterised by the existence of a cluster of mainly small and medium-sized enterprises in close geographical proximity (see among others, Becattini, 1979 and 1990; Balloni and Vicarelli, 1979; Cori and Cortesi, 1977);

- an *economic element*, characterised by an intense set of backward, forward and horizontal inter-firm linkages, based both on market and non-market exchanges of goods, services, information and human capital (see among others, Becattini, 1979 and 1990; Balloni and Vicarelli, 1979; Cori and Cortesi, 1977; Dei Ottati, 1986; Berardi and Romagnoli, 1984);

- a *socio-cultural element*, characterised by the relatively homogeneous cultural and social background linking the economic agents (see among others, Bagnasco, 1983 and 1988; Bagnasco and Trigilia, 1984, Bianchi, 1989; Fuà and Zacchia, 1983; Balloni and Vicarelli, 1979);

- an *institutional element*, characterised by a network of public and private local institutions supporting the economic agents in the cluster (see among others, Trigilia, 1985; Bagnasco and Trigilia, 1984).

According to this approach, locational advantage is secured by stable inter-SME linkages – both vertical and horizontal – and stable local labour markets.
In the 'industrial district model' the roles of inter-SME linkages are to:

- lower transaction costs by repeat contracts and the development of trust which reduces the need for costly search for partners and suppliers and for the formal specification of the terms of each economic transaction;

- generate *external* economies at the level of the district which reduce the cost disadvantage of small firms with respect to large firms. In particular, district economies cut the costs of information collection, vocational training, infrastructure and services.

The identification of dynamic benefits from customer-supplier relationships leads towards the concept of the *milieu*. As the GREMI group has underlined since the 1980s, inter-firm co-operation and tacit transfer of knowledge generate a collective learning process within the district which enhances the innovation capability of member firms.

An *innovative milieu* is defined as a set of relationships within a localised territory, encompassing in a coherent way a production system, different economic and social actors, a specific culture, a representational system, and generating a dynamic process of collective learning (Camagni, 1991). The *milieu innovateur* functions like a microcosm in which all those elements which are traditionally considered as the sources of economic development and change within the firm operate as if they were *in vitro*. This results from the high density of SME interactions, enhanced by spatial proximity and by those economic and cultural homogeneities which allow the milieu itself to exist. Essential components of the milieu innovateu include: a division of labour among units belonging to the same productive cycle; processes of learning-by-doing and learning-by-using, amplified beyond each enterprise by the high mobility of the specialised labour force inside the local area; external economies, generated by a common industrial culture and intense input-output interactions; entrepreneurship, facilitated by specific historical competences, sectoral specialisation and ample possibilities of imitation; and cross-fertilisation processes, which generate systems of integrated and incremental innovations.

Within the milieu, two kinds of co-operative processes are at work:

- a set of mainly informal, 'non-traded' relationships within the territory – among customers and suppliers and between private and public actors – and tacit transfers of knowledge by means of professional mobility and inter-firm imitation processes;

- more formalised co-operation agreements, mainly trans-territorial, with firms, collective agents, and public institutions in the field of technological development, vocational training, infrastructure and services provision.

The former kind of relationship is the 'glue' that creates a 'milieu' effect. This is complemented by the latter, more formal relationships, which can be interpreted as 'network relations' *proper*. Both sets of relationships may be regarded as tools or 'operators' that help small firms in their competitive struggle, reduce uncertainties intrinsic to the innovation processes and enhance their creativity (Camagni, 1991).

In the trans-territorial relations firms have with other enterprises, banks, research centres, training providers, local authorities and other institutions the locational dimension is only one of many co-ordinating mechanisms which serve to define the network. At a first glance, therefore, these networks are not invested with those milieu properties requiring close proximity. But they become more territorially based as they begin to generate localised development and identity (e.g., Apple at Cupertino, Silicon Valley), and when such network relations begin to multiply. In such cases, identification with the milieu often prevails over the identification with an individual partner, emphasising the importance of locality. For example, the strategic importance of links a firm has with a company in Silicon Valley resides more in the opening of a 'technological window' into Silicon Valley rather than simply accessing the know-how of a specific company.

From a theoretical perspective, the functions of the local milieu are twofold:

- firstly, it reduces the uncertainty inherent in innovation whilst minimising obstacles to change. In large companies, the uncertainty reducing function is carried out by information collection, its assessment and transcoding, selection of decision-making routines and control of competitors' decisions which are generally undertaken by R&D departments and/or strategic planning units. In the *milieu innovateur*, they are undertaken in a collective and socialised way by the milieu itself, through the rapid circulation of information and by the processes of imitation and co-operation (Camagni, 1991);

- secondly, the *milieu innovateur* supplies a durable substratum for learning processes and for guaranteeing the tacit transfer of know-how and non-codified immaterial assets among enterprises. In large enterprises, this continuity is guaranteed by the long term operation of R&D and engineering departments and by their continuous interactions. In small-firm clusters, where the life-cycle of single productive units is shorter, these functions are performed in a socialised way outside each enterprise and find their continuity in the labour market, in the local production culture and institutions, and in inter-personal and inter-generational links.

These two functions of the milieu are shown in Table 5.1, where they are classified by the two main, genetic characteristics of the local milieu:

- *geographical proximity*; and,

- *relational proximity* which encompasses the linkages that develop from the economic integration of firms, socio-cultural homogeneity, and dense public/private co-operation and partnership.

The fundamental characteristic of the processes resulting from geographical and relational proximity in the functioning of the milieu is that they are 'collective' rather than overtly co-operative, although additional benefit can be derived from firms working closely together. They operate within a competitive environment in which entrepreneurs are often highly individualistic rather than consciously co-operative. Moreover, the competitiveness of the milieu is not dependent on any explicit territorial marketing; the process of learning and accumulation of specific competencies is embedded in the local labour market rather than resulting from jointly organised training programmes; and imitation and other means of diffusing new technology is largely independent of the will of original innovators.

Inter-SME linkages are a vital element for the existence, functioning and survival of the milieu. They perform a double role, acting both as channels for the accumulation and exchange of knowledge – through collective learning mechanisms – and as means of reducing uncertainty. From this theoretical perspective, the role of inter-SME links is of crucial importance in innovative activities and performance. This claim is empirically tested in sections 5.3 and 5.4 which focuses on three main questions:[4]

- whether and to what extent do local SME linkages constitute channels for the construction, exchange and acquisition of local industrial knowledge;

- to what extent are local learning process exploited by firms in their search for a competitive advantage;

- to what extent do inter-SME links enhance the innovation process of firms?

5.3 The Evidence: Inter-SME Linkages and Learning Behaviour

The empirical analyses undertaken in different European high-technology

Table 5.1 Functions of the local milieu

		FUNCTIONS	
		Reduction of uncertainty	*Durable substratum for collective learning*
CONDITIONS	Geographical proximity	• Information collection • Vertical integration within 'filierès' • Local signalling (collective marketing)	• Labour turnover inside the milieu • Imitation of innovation practices
	Relational proximity	• Trust, achieved among partners • Co-operation in strategic decision-making • Risk sharing among partners • And, in some cases, mutual assistance and consensual sharing of the value chain	• Co-operation in industrial projects • Tacit transfer of knowledge • Public/private partnership in complex development schemes

milieux examined, among other things, the identification of the forms of inter-firm networking which characterise RTD-intensive SME clusters. The milieux included in the study are shown in Table 5.2. The main issues considered are: a) the density of local inter-SME relationships; b) the importance of these links in the innovative activities of firms; and c) the relative importance of learning processes inside the firms, those between firms within the local territory, and with firms outside the local area.

The analysis reveals *a relatively high proportion of firms reporting close links with other firms within the region*, especially those in vertical customer/supplier chains. In the Italian districts, there is a strong propensity to develop links with local customers and suppliers. Only 16 per cent of firms had suppliers outside the area and only 19 per cent had external customers (Campoccia, 1997; see Figure 5.1a and 5.1b).

A similar high density of local links was found in the Cambridge region. There, 76 per cent of firms (81 per cent in manufacturing) reported close links with other firms in the area (Table 5.3a). Among these links, those with suppliers (68 per cent) and with the providers of services are by far the most important whereas links with local firms in the same line of business, research collaborators, and customers were relatively unimportant (Table 5.3b).

124 *High Technology Clusters, Networking and Collective Learning in Europe*

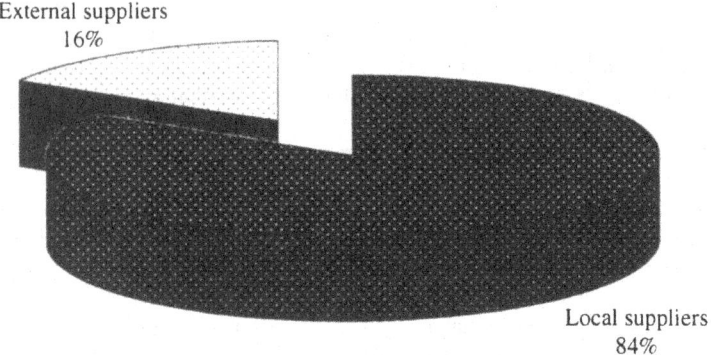

Figure 5.1a Italian high-technology SMEs: local and external suppliers

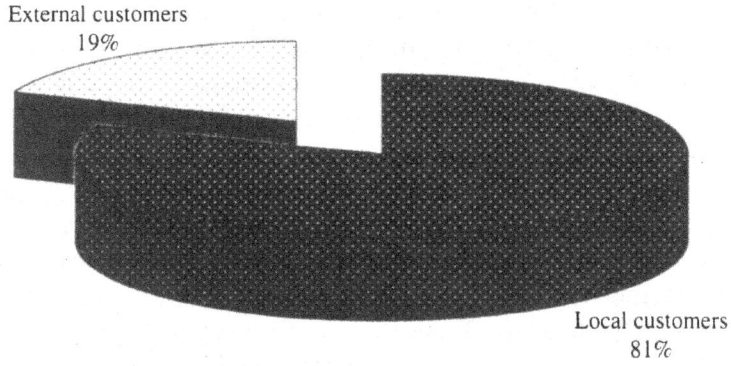

Figure 5.1b Italian high-technology SMEs: local and external customers

Source: Campoccia 1997.

Table 5.2 European high-technology milieux analysed

Country	Area	Sample size (number of interviewed firms)	Characteristics of the sample	Responsible team
United Kingdom	Cambridge region	50	Manufacturing and service firms	Keeble, Lawson, Moore, Wilkinson
United Kingdom	Oxfordshire	50	Manufacturing and service firms	Lawton Smith
Italy	Milan, Pisa Piacenza	63	Manufacturing and service firms	Camagni, Capello and Campoccia
The Netherlands	Utrecht region	55	Manufacturing and service firms	Wever
Germany	The Research Triangle region: Hannover, Brunswick, Göttingen	1,389	Manufacturing and service firms	Sternberg, Tamásy

Table 5.3a Local inter-firm links in the Cambridge region

	Number	%
Close inter-firm links with other firms in the region	38	76
of which manufacturing firms	17	81
Firms reporting that research links with other local firms had been important* for their development	20	40
of which manufacturing firms	11	52

Source: Lawson, 1997.

Table 5.3b The importance of local inter-firm links in the Cambridge region

	Number	%
Importance* of local links with:		
suppliers or subcontractors	26	68
firms providing services	20	53
customers	12	32
firms in your line of business	11	29
research collaborators	10	26

* Firms rating local link moderately, considerably or extremely important (3, 4 and 5) on a scale from 1 indicating completely unimportant to 5 indicating extremely important, as % of firms with links.

Source: Lawson, 1997.

In Oxfordshire, the proportion of firms having close local inter-SME links was 46 per cent. Nevertheless, the majority of high-technology firms (87 per cent) with close links with other local firms regarded these links as important or very important; and, furthermore, links with local suppliers within the region are recognised to be important or very important by more than 57 per cent of firms (Table 5.4).

Despite differences in the incidence of inter-firm links between the different milieux, evidence suggests that the links are important for the innovative activities of local firms. In the Cambridge region, 40 per cent of the firms reported that research links with other local firms had been important for their development. The share increases to 52 per cent when only manufacturing firms are analysed (Table 5.3a above). In Oxfordshire, where half of the firms report local links with other firms, more than 80 per cent estimate these links as important for their development (Table 5.4).

Table 5.4 The number and importance of different inter-firm links in Oxfordshire

	Number	%
Close local inter-firm links	23	46
Importance* of different types of links with:		
Customers	9	39
Suppliers/sub-contractors	13	57
Firms providing services	9	39
Research collaborators	9	39
Firms in your line of business	3	13
Total	20	87

* Firms rating link considerably or extremely important (4 and 5), as % of firms with links

Source: Lawton Smith, 1997.

Among all possible links, the majority of Oxfordshire firms rates customers or suppliers as the most important sources of innovation, although universities and government laboratories are also important sources of innovation (Table 5.5).

In Germany, the analysis carried out in the so-called research triangle Hannover-Brunswick-Göttingen confirms the importance of customers and suppliers as co-operative partners in innovative activities, findings which were independent of firm size. Customers and suppliers emerge (see Figure 5.2) as the second and third most important co-operation partners in the innovative process with business services occupying the first place (Sternberg and Tamásy, 1997).

Figure 5.3 (a and b) shows the importance of local suppliers and customers in the innovative activity of high-technology SMEs in the three Italian innovative milieux studied. 54 per cent and 68 per cent of firms respectively reported that links with suppliers and customers were important or highly important for their innovative activities. More detailed analysis revealed three clusters of Italian firms with different learning behaviour (Table 5.6; see also Capello, 1999).

- The first cluster is characterised by dynamic and innovative firms which enjoy traditional *industrial district* advantages of static efficiency

Table 5.5 Sources of innovation reported as moderately to highly significant by firms in Oxfordshire*

	Number	%
Within the region:		
suppliers of standardised components	5	11
suppliers of customised components	7	16
customers	12	26
competitors	5	11
consultancy firms	3	7
universities/government laboratories	15	32
Within the UK:		
suppliers of standardised components	13	31
suppliers of customised components	16	32
customers	33	69
competitors	16	27
consultancy firms	6	14
universities/government laboratories	23	51
Outside the UK:		
suppliers of standardised components	11	27
suppliers of customised components	14	33
customers	35	75
competitors	26	56
consultancy firms	3	7
universities/government laboratories	16	36

* Percentages are of number of firms responding to that question, not of the total sample

Source: Lawton Smith, 1997.

(industrial atmosphere and cultural proximity with the labour force). These locational benefits are added to by learning based primarily on know-how from outside the local area; external linkages which no doubt feed the innovative activities of the firms;

- The second cluster is a sub-system of autonomous firms within the milieu, for which learning is mainly based on internal competencies. Firms belonging to this cluster are specialised in process innovation and the main channels of learning, are, as expected, a) learning within the firm; b) technological proximity with customers and suppliers;

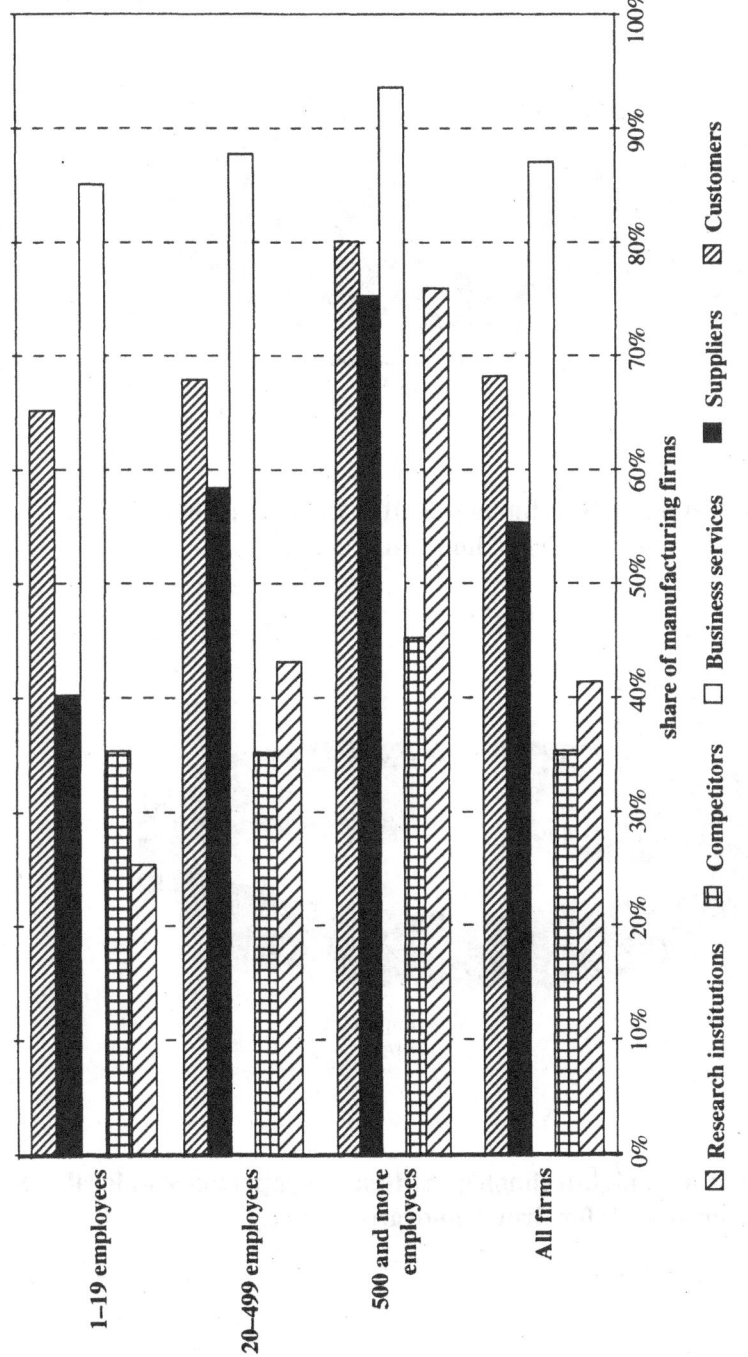

Figure 5.2 Innovative linkages of innovative manufacturing SMEs in the Hannover-Brunswick-Gottingen Research Triangle, by firm size and co-operation partner

Source: Sternberg and Tamásy, 1997.

130 *High Technology Clusters, Networking and Collective Learning in Europe*

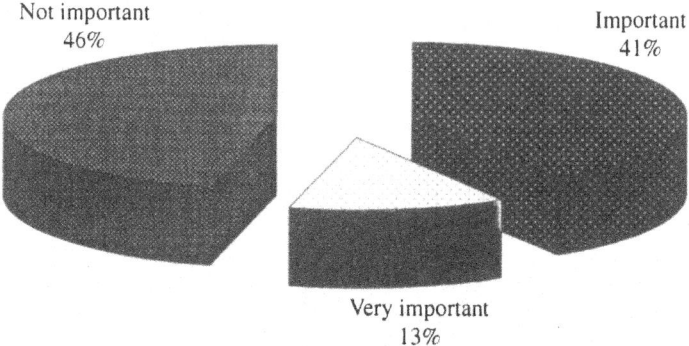

Figure 5.3a Italian high-technology SMEs: the importance of local suppliers for firms' innovative activity

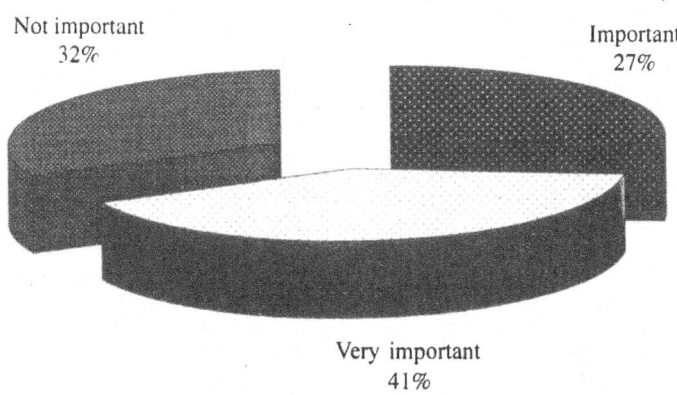

Figure 5.3b Italian high-technology SMEs: the importance of local customers for firms' innovative activity

Source: Campoccia, 1997.

- The third cluster reveals pure *milieu* behaviour where the learning process stems from socialised mechanisms of spatial transfer of knowledge, i.e. from collective learning. The smallest and most innovative firms producing radically new products belong to this group which is characterised by local spin-off, a high turnover of highly skilled workers within a stable local labour market, highly dynamic synergies with local suppliers with whom they have organisational and industrial proximity, and informal contracts with customers. These firms fully exploit local collective learning for their innovations.

In this analysis, an interesting result emerges concerning the customer/supplier relationships. Firms developing process innovation regard technological proximity with suppliers and customers as very important and common technical standards and technological backgrounds are facilitators of innovation. These elements are much less important for firms engaged in more complex product innovation when *institutional* and *organisational* proximity with both suppliers and customers become the crucial elements. Thus, cultural and social elements facilitate inter-SME links and appear to play a more important role in shaping innovative activity than do more traditional locational advantages. The presence of local innovative suppliers is also an important element for the most innovative firms, while presence of innovative customers appears to play a role in both process and product innovation (Table 5.6).

These findings are supported by the experience of firms in other milieux. In Oxfordshire, geographical proximity to suppliers and customers turned out not to be important for firms' development; only 18 per cent of the firms underline spatial proximity with suppliers as an important source for their development. The proportion becomes even less (10 per cent) for the importance of geographical proximity with customers (Table 5.7). In Cambridge, local links with suppliers are important for over half of all high-technology SMEs (Table 5.3b: see also Figure 5.4). However, geographical proximity to and links with local customers are rated less highly (Figure 5.4).

In the Utrecht region, while relationships with suppliers emerge as important sources of innovation for 74 per cent of firms and with customers for more than 90 per cent, geographical proximity plays a much less important role. Only 21 per cent of interviewed firms reported that their relationship with regional customers was important, and only 36 per cent recognise links with suppliers as significant (Figures 5.5 and 5.6). The relational space of Dutch high-tech firms is not regional, rather it is the

Table 5.6 Results of a cluster analysis of learning behaviour in the three Italian milieux

Factors	Cluster 1	Cluster 2	Cluster 3
Dynamic and innovative firms	**0.127**	-1.32	-0.24
Industrial atmosphere	**0.047**	-0.33	-0.48
Cultural proximity	**0.010**	-0.039	-0.20
Learning external to the miliex	**0.02**	-0.19	-0.18
Process innovative firms	-0.01	**0.25**	-0.103
Proximity to the mother firm	0.009	**0.08**	-0.46
Learning internal to the firm	-0.04	**0.55**	-0.18
Presence of local innovative customers	-0.05	**0.51**	0.28
Technological proximity with customers	0.039	**0.17**	-1.53
Technological proximity with suppliers	0.009	**0.15**	-0.64
Smallest and most innovative firms	-0.0009	-0.35	**0.89**
Market stability	0.014	-0.30	**0.33**
High labour force turnover	0.0021	-0.18	**0.41**
Spin-off	0.0027	-0.039	**0.02**
Standard contracts with customers	0.02	-0.52	**0.63**
Institutional and organisational proximity with customers	-0.05	0.29	**0.68**
Presence of local innovative suppliers	-0.059	0.42	**0.61**
Institutional and organisational proximity with suppliers	-0.03	0.18	**0.47**

Note: values in the table represent the average value of each variable for each group of firms; values in bold indicate main variables characterising each cluster.

Source: Capello, 1999.

Table 5.7 The importance* of local links for firms' development in Oxfordshire

Type of links	Number	%
Research links with other Oxford firms	16	32
Proximity to local customers	9	18
Proximity to local suppliers	5	10
Access to local business services	5	10

*   Firms rating local link moderately, considerably or extremely important (3, 4 and 5) on a scale from 1 indicating completely unimportant to 5 indicating extremely important

Source: Lawton Smith, 1997.

The Role of Inter-SME Networking and Links in Innovative High-Technology Milieux 133

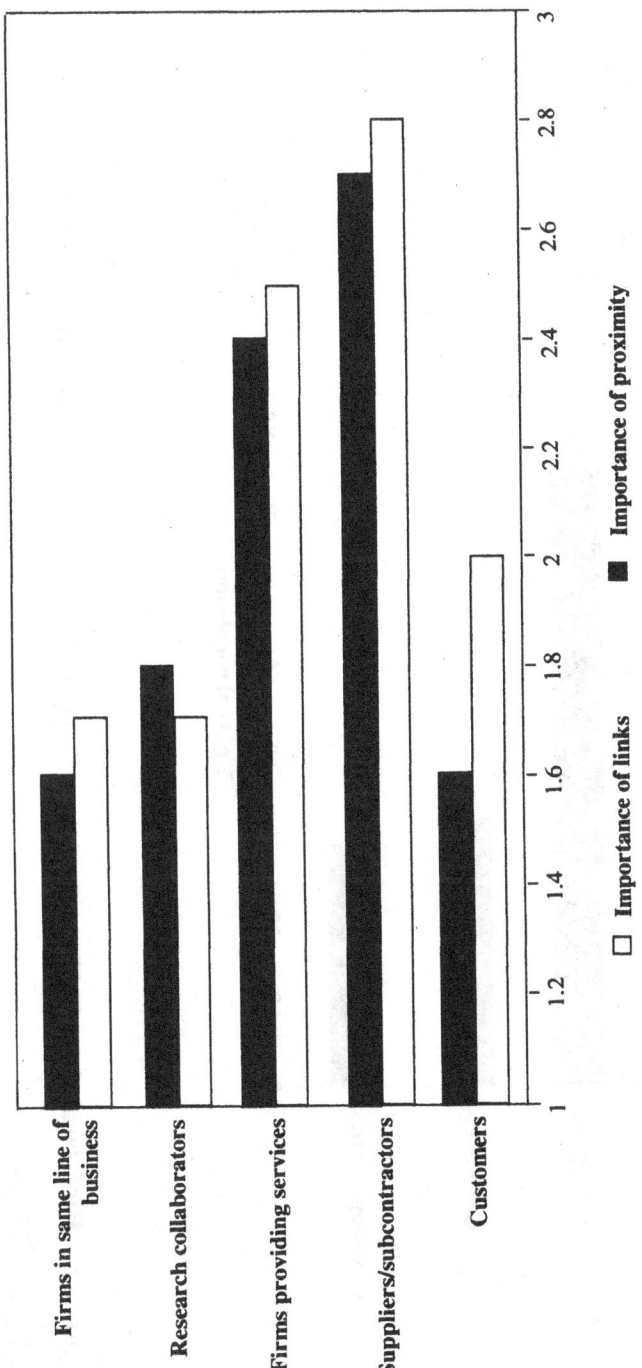

Note: the chart shows mean values of firm responses on a scale from 1 (insignificant) to 5 (highly significant).

Figure 5.4 The importance of local inter-firm links to Cambridge technology-intensive SMEs

Source: Lawson, 1997.

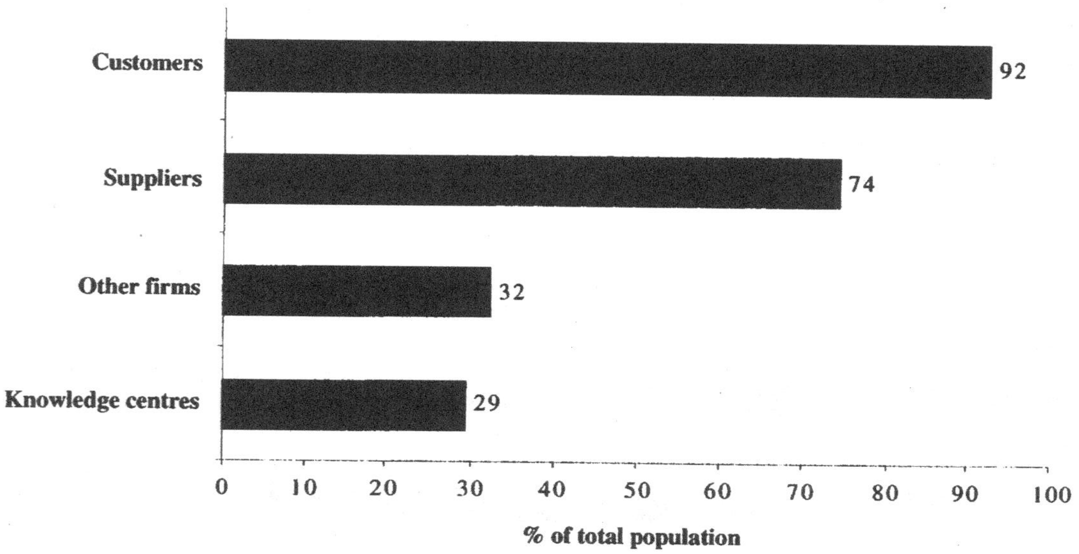

134 *High Technology Clusters, Networking and Collective Learning in Europe*

Figure 5.5 High-technology SMEs in the Utrecht region: the importance of innovation partners

Partner	% of total population
Customers	92
Suppliers	74
Other firms	32
Knowledge centres	29

Source: Wever, 1997.

The Role of Inter-SME Networking and Links in Innovative High-Technology Milieux 135

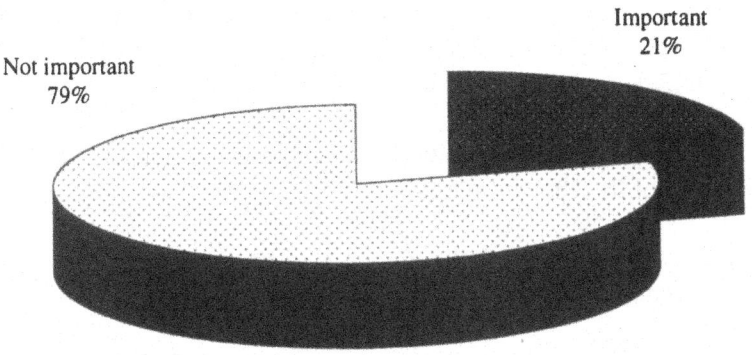

Figure 5.6a High-technology SMEs in the Utrecht region: the importance of regional customers

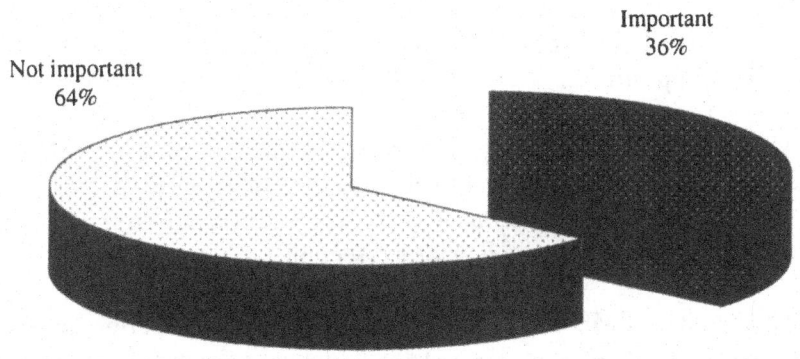

Figure 5.6b High-technology SMEs in the Utrecht region: the importance of regional suppliers

Source: Wever, 1997.

national level which is characterised by a high internal integration (Wever, 1997). There is no evidence that organisational and cultural proximity plays any part in the innovation process in the Netherlands.

Despite differences in the regional contexts and in the size of the samples some common tendencies can be observed in the local linkages between high-technology SMEs:

- a high incidence of links with local suppliers and customers, with the exception of the Utrecht region;

- a general recognition of the importance attributed to such links for internal innovation processes of these firms;

- the importance of vertical links does not stem from geographical proximity, but rather from cultural, social and institutional similarities.

5.4 The Evidence: Inter-SME Linkages and Innovation Processes

Previous sections have provided a descriptive account of the density of inter-SME links, and suggested reasons for their development. In this section, we present a more interpretative analysis of the role inter-SME links play in enhancing the innovative activity of firms. This analysis is based on data from the three Italian milieux (Pisa, Piacenza and Milan) and will test the following propositions:

- whether inter-SME links, especially between customers and suppliers, are channels for the acquisition of knowledge. If so, *firms exploiting intense interactions with local customers and suppliers should display a greater labour productivity than other firms* (section 5.4.2);

- whether inter-SME links are ways of achieving a superior innovative capability in local areas. If so, we would expect *product innovation and breakthrough innovation to be positively correlated with relationships with customers and suppliers* (section 5.4.3).

5.4.1 Sample and Data

In order to test the theoretical propositions presented above, high-technology firms were surveyed in three high-technology milieux in Italy – Pisa, Piacenza and the North-Eastern part of Milan. Information collected included:

- the characteristics of the firms, in terms of employees, turnover, innovative activity, economic dynamics;

- the characteristics of the local labour market, in terms of quality of the labour force, formal and informal channels for labour recruitment, the turnover within the firm, stability of the labour market;

- the relationships with customers and suppliers, emphasising the role of local linkages and of spatial, as well as institutional and organisational, proximities among them.

Some 63 firms in the three milieux were interviewed and a database was constructed consisting of binary (yes, no) and discrete (qualitative judgement) variables.

Factor analysis, which summarises a large number of variables with a smaller number of 'derived' variables, was the method used to address the above theoretical issues. The full results of the factor analysis are presented in Appendix 5.1.

5.4.2 Inter-SME links and Factor Productivity

In this section we address the first testable hypothesis by comparing the factor productivity of firms who attribute different degrees of importance to vertical linkages. One of the main implications of the existence of inter-SME links is that firms in the milieu who regard geographical, organisational and institutional[5] proximity with suppliers and customers as important for their innovative activity should, other things being equal, have higher levels of factor productivity than firms for which such linkages are less important.

A production function is estimated for firms, where both capital and labour parameters depend on both the presence of innovative customers in the area, and institutional and organisational proximity with customers. In particular, the model tests how labour and capital productivity changes with regard to two indicators representing the presence of innovative customers in the area, and proximity to them in terms of organisational and institutional aspects[6] (see Appendix 5.2 for the mathematical and econometric details). For this purpose the relevant capital is 'non-material capital', i.e. the internal knowledge of the firm measured by R&D expenditure and patents.[7]

As far as labour productivity is concerned, the results obtained from the estimation of the production function model are satisfactory from a statistical point of view, with an R square of 0.63 (see Table 5.10 in Appendix 5.2 for the statistical results) From an interpretative point of view, the results are also rather interesting and provide support for the theoretical hypotheses. Marginal productivity of labour increases at decreasing rates (Figure 5.7a). This situation changes when, *ceteris paribus*, labour productivity is measured for different levels of local customers' links. In particular, firms rating the presence of local innovative customers as more important register greater labour productivity (Figure 5.7b).

The tendency to increasing labour productivity becomes more pronounced when it is correlated with institutional and organisational proximity with customers; firms giving a high importance to these kinds of advantages are also those showing greater labour productivity (Figure 5.7c).

These results indirectly support the hypothesis that local inter-SME links are a channel for knowledge acquisition. Moreover, the role played by physical proximity seems to be positive but less important than those played by institutional and organisational proximity. This reinforces the main findings of section 5.3.

As far as capital productivity is concerned, the results are satisfactory in terms of economic interpretation, although their statistical significance as measured by Student's T values are low (see Table 5.10 in Appendix 5.2). The validity of the overall model is in any case witnessed by a high R square value (0.63), as mentioned above. Non-material capital shows a U shape, suggesting that a critical mass of knowledge is necessary to turn innovative knowledge into tangible results (Figure 5.7d).

Figure 5.7e shows what happens to non-material capital productivity when firms rate proximity with innovative customers more. In such firms, capital productivity decreases at increasing rates, in contrast with the expected results (Figure 5.7e). Institutional and organisational proximity with customers, in contrast, generates diminishing capital productivity, but at a decreasing rate (Figure 5.7f); and at higher levels capital productivity increases again. Once again, these results suggests that what really makes the difference is institutional and organisational, rather than simply geographical, proximity.

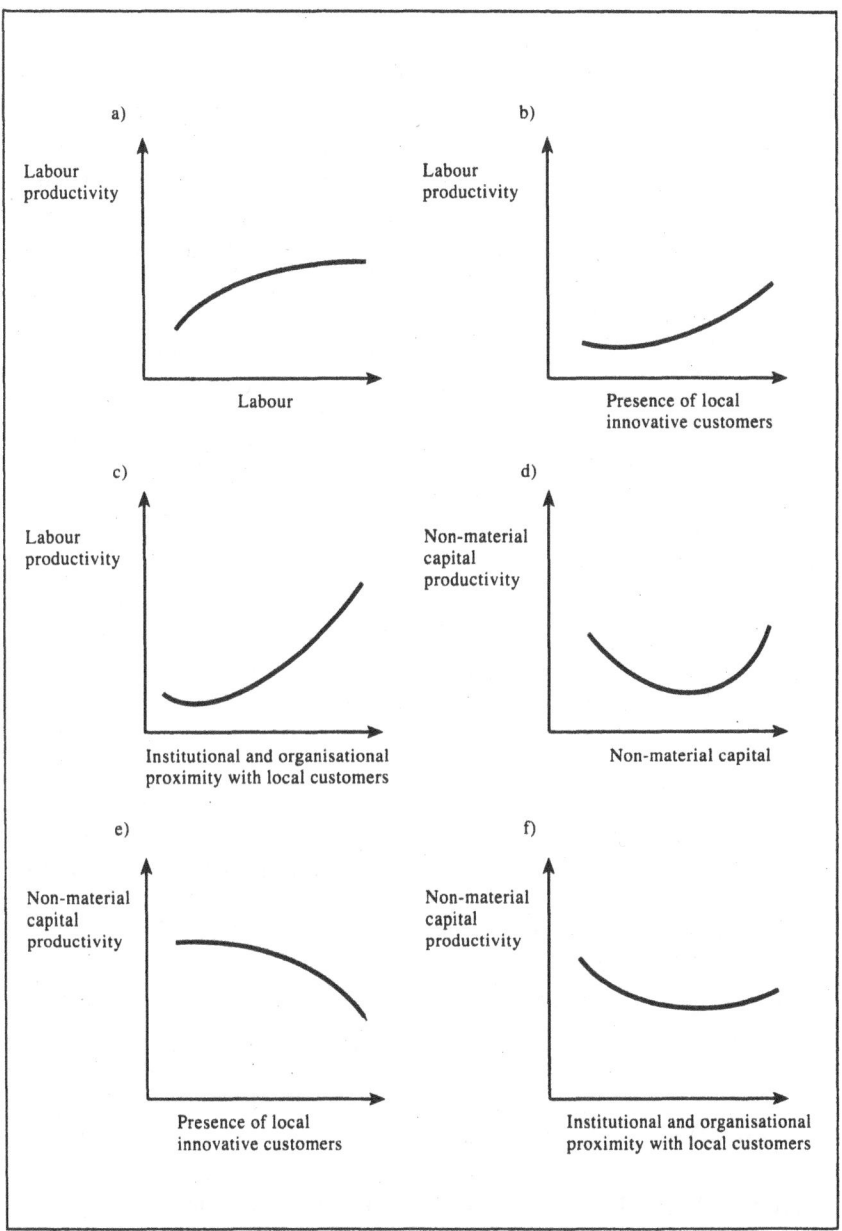

Figure 5.7 Factor productivity with respect to local inter-SME links

5.4.3 Inter-SME links and Innovative Activities

Our second hypothesis relates to the role played by inter-SME links in the innovative activities of firms in the milieux. If links with other firms are ways of securing new creative resources for SMEs and a tool for developing a localised pool of knowledge, we would expect product innovation, and particularly radical innovation, to be positively correlated with inter-SME links.

This hypothesis was tested using linear regression in which the three variables characterising the structural and innovative characteristics of the sample firms (DININ, SMIN and PROCIN – see Appendix 5.1 for details) are the dependent variables, and are regressed on the other variables obtained from the factor analysis. The results confirm our expectations and our hypotheses (Table 5.8):

- the radical innovation activity of the smallest firms (SMIN) depends on the presence of innovative customers in the area. This variable is significant from a statistical point of view, and has the expected positive sign;

- process innovation activity (PROCIN) is negatively correlated with institutional and organisational proximity with customers, suggesting an independence of process innovation activities from elements of dynamic synergy.

Table 5.8 Results from regression analyses

Independent variables	Dependent variables: SMIN	PROCIN
Presence of innovative customers	0.488 (1.90)	
Institutional and cultural proximity with customers		-0.14 (-2.14)

5.5 Conclusions

The purpose of this chapter is to provide an insight into the nature of inter-SME linkages in high-technology milieux, and their effects on the innovative activity and productivity of local firms from both a theoretical and empirical standpoint.

In both the literature on 'industrial districts' and the more recent on *milieux innovateurs* an important role is attributed to inter-SME links, and especially vertical relationships. However, in the two theoretical approaches differences exist in the role attributed to supplier-customer linkages: in industrial districts theory they contribute to the achievement of local competitive advantages, in that:

- they help in generating external economies, by providing collective efficiency and industrial atmosphere *à la* Marshall;

- they help in reducing transaction costs, thanks to geographical, organisational, and social proximity among economic agents.

In this approach, static in nature, inter-SME linkages play a vital role in creating local competitive advantage.

The GREMI *milieux innovateurs* approach provides a dynamic interpretation of customer-supplier relationships based on the economic importance of cooperation and continuity. In the local milieu, inter-SME links provide: a) channels for knowledge acquisition; b) channels for uncertainty reduction; and c) channels of a local collective learning. The empirical analysis undertaken by the European network provides empirical evidence supporting the existence and the importance of these inter-SME links in the innovative process.

The chapter has also presented interpretative results on the role played by vertical inter-SME links in factor productivity: labour productivity increases with institutional and organisational proximity to customers.

Some general concluding remarks can be drawn from the analysis:

- inter-SME links with suppliers and customers are present in these areas;

- inter-SME links are recognised by firms to be important elements in their innovative activities. This result confirms a previous empirical finding by the ESRC Centre for Business Research, namely that high-technology SMEs exhibit significantly higher levels of inter-firm networking and collaboration than less technologically-focused SMEs (Keeble et al., 1997);

- the quality of inter-SME links has a positive effect on labour productivity;

- the choice of local suppliers and customers is linked more to institutional and organisational proximity than to geographical proximity.

However, different patterns and paths to knowledge acquisition emerge in high-technology districts: through network linkages with external firms, through the full exploitation of the internal knowledge of multi-localised firms, and through the collective learning process encouraged by the local milieu and labour market.

Notes

1 Though the chapter is the result of joint research, R. Camagni has written section 5.2, while the remaining sections have been written by R. Capello.
2 On local districts theory see, among others, Bagnasco and Trigilia, 1984; Becattini, 1979 and 1990; Bianchi, 1989; Brusco, 1982. For a review, see Bramanti and Maggioni, 1997.
3 On the concept of *milieu innovateur*, developed by the international scientific association called GREMI, see, among others, Aydalot, 1986; Aydalot and Keeble, 1988; Camagni, 1991; Maillat et al., 1993; Ratti et al., 1997.
4 The empirical analysis is based on work carried out by the TSER European network led by the ESRC Centre for Business Research of the University of Cambridge.
5 By institutional proximity a situation is meant in which local firms sign contracts with suppliers and customers on the basis of the same legal rules, while by organisational proximity a situation is meant in which firms work on the basis of the same organisational routines and organisational structures.
6 The model is estimated taking into consideration links with customers. In contrast, links with suppliers do not lead to unacceptable statistical results.
7 These data were obtained from the balance sheet of firms.

References

Aydalot, Ph. (ed.) (1986), *Milieux Innovateurs en Europe*, GREMI, Paris.
Aydalot, Ph. and Keeble, D. (eds) (1988), *High Technology Industry and Innovative Environments*, Routledge, London.
Bagnasco, A. (1983), 'Il contesto sociale', in G. Fuà and C. Zacchia (eds), *Industrializzazione Senza Fratture*, Il Mulino, Bologna.
Bagnasco, A. (1988), *La Costruzione Sociale del Mercato*, il Mulino, Bologna.
Bagnasco, A. and Trigilia, C. (1984), *Società e Politica nelle aree di Piccola Impresa: Il Caso di Bassano*, Arsenale, Venezia.
Balloni, V. and Vicarelli, R. (1979), 'Il Sistema Industriale Marchigiano Negli Anni Settanta', *Economia Marche*, no. 5, pp. 43–74.
Becattini, G. (1979), 'Dal settore Industriale al Distretto Industriale: alcune Considerazioni sull'Unità di Indagine della Politica Industriale', *Economia e Politica Industriale*, no. 1, pp. 1–79.

Becattini, G. (1990), 'The Marshallian Industrial District as a Socio-economic Notion', in F. Pyke, G. Becattini and W. Sengenberger (eds), *Industrial Districts and Inter-firm Cooperation in Italy*, ILO, Geneva.

Berardi, D. and Romagnoli, M. (1984), *L'Area Pratese tra Crisi e Mutamento*, Consorzio Centro Studi, Prato.

Bianchi, P. (1989), 'Concorrenza Dinamica, Distretti Industriali e Interventi Locali', in F. Gobbo (ed.), *Distretti e Sistemi Produttivi alle Soglie degli Anni Novanta*, Franco Angeli Editore, Milan, pp. 47–60.

Bramanti, A. and Maggioni, M. (eds) (1997), *La Dinamica dei Sistemi Produttivi Territoriali: Teorie, Tecniche, Politiche*, Franco Angeli Editore, Milan.

Brusco, S. (1982), 'The Italian Model: Productive Decentralisation and Social Integration', *The Cambridge Journal of Economics*, vol. 6, pp. 167–84.

Camagni, R. (1991), '"Local Milieu", Uncertainty and Innovation Networks: Towards a New Dynamic Theory of Economic Space', in R. Camagni (ed.), *Innovation Networks: Spatial Perspectives*, Belhaven, London, pp. 121–44.

Campoccia, A. (1997), 'Inter-SMEs Linkages in Italian High-Technology Milieux: Empirical Evidence', in D. Keeble and C. Lawson (eds), *Networks, Links and Large Firm Impacts on the Evolution of Regional Clusters of High-Technology SMEs in Europe*, ESRC Centre for Business Research, University of Cambridge, Cambridge, pp. 25–40.

Capello, R. (1999), 'Spatial Transfer of Knowledge in High-Technology Milieux: Learning Versus Collective Learning Processes', *Regional Studies*, vol. 33, pp. 353–65.

Chambers, R.G. (1988), *Applied Production Analysis*, Cambridge University Press, Cambridge.

Christensen, L., Jorgenson, D. and Lau, L. (1973), 'Transcendental Logarithmic Production Frontiers', *The Review of Economics and Statistics*, vol. 55, pp. 28–43.

Cori, B. and Cortesi, G. (1977), *Prato: Frammentazione ed Integrazione di un Bacino Tessile*, Fondazione Giovanni Agnelli Editore, Turin.

Dei Ottati, G. (1986), 'Distretto Industriale, Problemi delle Transazioni e Mercato Comunitario: Prime Considerazioni', *Economia e Politica Industriale*, no. 51, pp. 93–121.

Evans, D. and Heckman, J. (1984), 'A Test of Suadditivity of the Cost Function with Application to the Bell System', *American Economic Review*, vol. 74, pp. 615–23.

Fuà, G. and Zacchia, C. (eds), *Industrializzazione Senza Fratture*, Il Mulino, Bologna.

Henderson, J.V. (1974), 'The Sizes and Types of Cities', *The American Economic Review*, vol. LXIV, pp. 640–56.

Kawashima, T. (1975), 'Urban Agglomeration Economies in Manufacturing Industry', *Papers of the Regional Science Association*, vol. 34, pp. 157–175.

Keeble, D., Lawson, C., Lawton Smith, H., Moore, B. and Wilkinson, F. (1997), 'Internationalisation Processes, Networking and Local Embeddedness in Technology-Intensive Small Firms', in M. Ram, D. Deakins and D. Smallbone, D. (eds), *Small Firms: Enterprising Futures*, Paul Chapman, London, pp. 60–72.

Lawson, C. (1997), 'Local Inter-Firm Networking by High-Technology Firms in the Cambridge Region', in D. Keeble and C. Lawson (eds), *Networks, Links and Large Firm Impacts on the Evolution of Regional Clusters of High-Technology SMEs in Europe*, ESRC Centre for Business Research, University of Cambridge, Cambridge, pp. 41–56.

Lawton Smith, H. (1997), 'Inter-Firm Networks in Oxfordshire', in D. Keeble and C. Lawson (eds), *Networks, Links and Large Firm Impacts on the Evolution of Regional Clusters of*

High-Technology SMEs in Europe, ESRC Centre for Business Research, University of Cambridge, Cambridge, pp. 57–66.
Maillat, D., Quévit, M. and Senn, L. (1993), *Réseaux d'Innovation et Milieux Innovateurs: un Pari pour le Développement Régional*, EDES, Neuchâtel.
Marelli, E. (1981), 'Optimal City Size, the Productivity of Cities and Urban Production Functions', *Sistemi Urbani*, 1/2, pp. 149–63.
Rabellotti, R. (1997), *External Economies and Cooperation in Industrial Districts: A Comparison of Italy and Mexico*, Macmillan, London.
Ratti, R., Bramanti, A. and Gordon, R. (eds) (1997), *The Dynamics of Innovative Regions*, Ashgate, Aldershot.
Segal, D. (1976), 'Are there Returns to Scale in City Size?', *The Review of Economics and Statistics*, vol. LVIII, pp. 339–50.
Shefer, D. (1973), 'Localization Economies in SMSA's: a Production Function Analysis', *Journal of Regional Science*, vol. 13, pp. 55–64.
Sternberg, R. and Tamásy, C. (1997), 'SMEs and Large Firms in Germany: How do They Differ in Terms of Innovative Linkages?', in D. Keeble and C. Lawson (eds), *Networks, Links and Large Firm Impacts on the Evolution of Regional Clusters of High-Technology SMEs in Europe*, ESRC Centre for Business Research, University of Cambridge, Cambridge, pp. 103–16.
Sveikauskas, L. (1975), 'The Productivity of Cities', *Quarterly Journal of Economics*, vol. 89, pp. 393–413.
Trigilia, C. (1985), 'La Regolazione Logistica: Economia e Politica Nelle Aree di Piccola Impresa', *Stato e Mercato*, vol. 14, pp. 181–228.
Wever, E. (1997), 'Clusters of High-technology SMEs in a Small Homogeneous Country: The Netherlands', in D. Keeble and C. Lawson (eds), *Networks, Links and Large Firm Impacts on the Evolution of Regional Clusters of High-Technology SMEs in Europe*, ESRC Centre for Business Research, University of Cambridge, Cambridge, pp. 77–92.

Appendix 5.1 Factor and cluster analyses

The methodology used to describe learning behaviour for innovative activities among our firm sample is a cluster analysis. However, before entering the behavioural analysis, factor analysis is run, with the primary goal of simplifying the description of local systems and of their innovative and learning behaviour. Factor analysis allows the identification of a relatively small number of underlying principal elements of 'factors' that explain the correlations among a set of variables; in other words, it summarises a large number of variables with a smaller number of 'derived' variables. In fact, from our questionnaire, many variables could be used to describe:

a) firm characteristics, in terms of:

- growth, size, and innovative activity;
- relationships with suppliers, in terms of the role played by suppliers in the innovative activity of the firm, and whether organisational and institutional proximity matters;
- relationships with customers, as in the case of suppliers;

b) the local area characteristics, in terms of:

- district locational advantages, like industrial atmosphere, stable labour market, cultural proximity with the labour force;
- local labour market, especially in terms of inter-SME links with suppliers and customers, and especially the role played by institutional, organisational and geographical proximity to customers and suppliers.

Factor analysis has been run in order to identify for each group of characteristics mentioned above, which could be represented by many explanatory variables of our questionnaire, a smaller number of 'derived' variables. In statistical terms, the results are quite satisfactory: all factor analyses run in each of the above mentioned groups of characteristics explain a large share of total sample variance (Table 5.9 presents the results).

This statistical exercise is needed to run a multivariate cluster analysis, based on the factors identified, instead of the original variables. Cluster analysis is the methodology used to identify different learning behaviours with respect to different innovative activities and goals of firms. In statistical terms, cluster analysis groups firms according to their degree of similarity or proximity with respect to the main underlying factors which characterise the economic structure and the local relationships of the sample (Rabellotti, 1997).

Table 5.9a Factor analysis of the structural and innovative characteristics of firms

Variables	DININ	SMIN	PROCIN
Turnover over the sample average	0.51	-0.74	0.01
Increasing turnover	0.74	0.04	0.18
75% of the turnover depending on innovation	0.35	0.84	-0.04
Significant product innovation developed over the last 5 years	0.45	0.08	-0.61
Breakthrough product innovation developed over the last 5 years	0.64	0.012	-0.15
Significant process innovation developed over the last 5 years	0.19	0.009	0.84
Explained variance by each factor (in %)	26	21	19
Share of total explained variance: 67%			

Note: values in the table represent the factor loading of each variable on each of the factors.

Source: Capello, 1999.

Table 5.9b Factor analysis of suppliers' relationships

Variables	TECHPROS	ISTORGPROS	PRELOCS
Standard supply contract	0.43	0.63	-0.46
Contract based on technical standard	0.62	-0.1	-0.1
More than 75% of suppliers are local	0.08	0.22	0.85
Very important role played by suppliers in technical innovative processes	0.64	-0.0002	0.44
Common approach with suppliers to institutional aspects	0.002	0.83	0.18
Common approach with suppliers to technical aspects	0.79	0.17	0.16
Common approach with suppliers to organisational aspects	0.21	0.8	0.24
Complementary knowledge	0.87	0.14	0.08
High trustworthiness in cooperation	0.82	0.17	0.01
Explained variance by each factor (in %)	34	21	14
Share of total explained variance: 69%			

Note: values in the table represent the factor loading of each variable on each of the factors.

Source: Capello, 1999.

Table 5.9c Factor analysis of customers' relationships

Variables	ISTORGPROS	TECHPROC	PRELOCC	STANDCONC
Standard customer contract	0.27	-0.63	-0.07	0.56
Contract based on technical standard	-0.1	0.11	0.007	0.87
More than 75% of customers are local	0.13	0.05	0.84	-0.30
Very important role played by customers in technical innovative processes	0.18	0.11	0.83	0.28
Common approach with customers to institutional aspects	0.89	0.23	0.24	-0.05
Common approach with customers to technical aspects	0.37	0.79	0.09	0.023
Common approach with customers to organisational aspects	0.92	0.28	0.14	-0.007
Complementary knowledge	0.41	0.75	0.04	0.14
High trustworthiness in cooperation	0.30	0.5	0.37	0.32
Explained variance by each factor(in %)	24	22	18	15
Share of total explained variance: 79%				

Note: values in the table represent the factor loading of each variable on each of the factors.

Source: Capello, 1999.

Table 5.9d Factor analysis of location advantages

Variables	CULTPRO	ORFIRPRO	MKTSTAB	INDATM
Proximity to motorways and airports	-0.02	-0.63	0.27	0.06
Cultural and industrial atmosphere	0.29	0.26	-0.06	0.84
Lower production costs	-0.12	0.35	-0.64	-0.02
Common culture	0.81	-0.02	0.15	-0.10
Common technical background	0.81	0.03	-0.27	0.21
Stable local labour force	-0.11	0.08	0.80	-0.07
Very important role played by the local labour market in providing high quality labour force	0.49	0.18	0.44	0.17
Proximity to the residential place	0.34	0.50	-0.06	-0.64
Proximity to the original firm	0.03	0.81	0.20	0.19
Explained variance by each factor (in %)	20	17	16	14
Share of total explained variance: 67%				

Note: values in the table represent the factor loading of each variable on each of the factors.

Source: Capello, 1999.

Table 5.9e Channels of knowledge acquisition

Variables	LEXDIS	INLEAR	TURN	SPIN
The firm is the result of a spin-off	-0.12	-0.06	-0.15	0.74
Firm's technicians and scientists were previously employed in firms outside the milieu	0.54	-0.41	-0.26	-0.25
Firm's technicians and scientists were previously employed in local firms	-0.73	-0.29	-0.09	0.31
Firm's technicians and scientists were previously employed in local research centres	0.47	0.39	0.05	0.46
Firm's technicians and scientists were previously employed in external research centres	0.02	0.09	-0.12	-0.62
Firm's technicians and scientists had their training within the firm	0.21	0.75	-0.10	-0.23
Firm's technicians and scientists had their training outside the firm	-0.13	-0.85	-0.17	0.005
More than 50% of firm's labour force has been recruited by the firm in the last 5 years	0.19	0.28	0.81	0.18
More than 50% of firm's labour force has left the firm in the last 5 years	-0.11	-0.11	0.89	-0.15
Importance of recruiting technicians through informal channels	0.51	0.08	0.02	0.11
Importance of recruiting scientists through informal channels	0.75	0.11	0.01	-0.07
Explained variance by each factor (in %)	18	16	15	13
Share of total explained variable: 62%				

Note: values in the table represent the factor loading of each variable on each of the factors.

Source: Capello, 1999.

Appendix 5.2 The production function model

The methodology applied to this study is similar to the one applied during the seventies to estimate agglomeration economies at an urban aggregate[1] (urbanisation economies) or sectoral[2] (localisation economies) level. In general, these models applied a Cobb-Douglas production function, which allows the simple least-square method for the estimate of its parameters. These studies were able to estimate the effects of urban size on factor productivity, and thus agglomeration advantages obtained from large scale cities.

A similar methodology is applied to this study, with the exception that we choose a Trascendental Logarithmic function (Translog), which, contrary to the Cobb-Douglas production function, allows us to estimate second order variables and cross-variable effects[3] (Christensen et al., 1973). Our firm production function thus becomes:

$$\ln Y = \ln \eta + a\ln K + b\ln L + c\tfrac{1}{2}\ln K^2 + d\tfrac{1}{2}\ln L^2 + f \ln K \ln L \quad (1)$$

where Y is the firm's outcome, and K and L are respectively capital and labour. If the firm is located in a milieu, factor productivity is influenced by inter-SME links, being one of the ways for exploiting collective learning and thus spatial transfer of knowledge; these theoretical assumptions may be formalised as follows:

$$a = a_0 + a_1 CP + a_2 CIO \quad (2)$$

where a is the capital efficiency parameter which may depend on the presence of innovative customers in the area (CP), and by the institutional and organisational proximity with customers (CIO), as claimed above. The same can be said for labour, and its efficiency parameter reads as:

$$b = b_0 + b_1 CP + b_2 CIO \quad (3)$$

If we substitute in equation (1) equations (2) and (3), we obtain:

$$\ln Y = \ln \eta + a_0 \ln K + a_1 KH \ln K + a_2 EL \ln K + b_0 \ln L + b_1 SMK \ln L + d\tfrac{1}{2}\ln L^2 + f \ln L \ln K \quad (4)$$

This kind of function allows the direct estimate of outcome elasticity with respect to all input variables, capital and labour, that is the percentage outcome changes due to a 1 percent change of a specific determinant, other things being equal. In order to test whether the change in the capital has increased or decreased the firm's outcome (*ceteris paribus*) it is enough to calculate the following expression and to test the sign of e_K:

$$e_K = a_0 + a_1 CP + a_2 CIO + c \ln K + f \ln L \tag{5}$$

where e_K represents the outcome elasticity with respect to capital. At the same time, outcome elasticity with respect to labour is easily obtained, by calculating the following expression and testing the sign of e_L:

$$e_L = b_0 + b_1 CP + b_2 CIO + d \ln K + f \ln L \tag{6}$$

Moreover, thanks to equation 5 we can estimate how outcome elasticity with respect to capital changes when the rating given to the presence of local innovative customers changes, by simply calculating:

$$e_{KCP} = a_1 \tag{7}$$

and how outcome elasticity with respect to capital changes when institutional and organisational proximity to customers is rated differently, by calculating:

$$e_{KCIO} = a_2 \tag{8}$$

The same reasoning can be applied to outcome elasticity with respect to labour, by measuring how it changes, for different levels of presence of customers and by institutional and organisational proximity to customers, by calculating respectively:

$$e_{LCP} = b_1 \tag{9}$$
$$e_{LCIO} = b_2 \tag{10}$$

Table 5.10 reports the econometric results.

Table 5.10 Results of the econometric estimation

	Coefficient	T-student
Labour productivity		
– in firms having an average level of labour	0.943	5.56
– in firms having the highest level of labour	1.15	1.53
– in firms having the lowest level of labour	0.784	1.24
– in firms exploiting the presence of innovative customers less	0.943	5.56
– in firms exploiting the presence of innovative customers more	0.943	5.57
– in firms exploiting institutional and org. proximity to customers less	0.942	5.64
– in firms exploiting institutional and org. proximity to customers more	0.944	5.48
	Coefficient	T-student
Non-material capital productivity		
– in firms having an average level of non-material capital	-0.009	-0.06
– in firms having the highest level of non-material capital	-0.43	-1.31
– in firms having the lowest level of non-material capital	0.19	0.55
– in firms exploiting the presence of innovative customers less	-0.0098	-0.062
– in firms exploiting the presence of innovative customers more	-0.01	-0.06
– in firms exploiting institutional and org. proximity to customers less	-0.013	-0.08
– in firms exploiting institutional and org. proximity to customers more	0.006	0.04

Notes

1. See, among others, Henderson, 1974; Marelli, 1981; Segal, 1976; Sveikauskas, 1975.
2. Sectoral analyses are contained in Kawashima, 1975; Shefer, 1973.
3. A major debate exists around the necessity to impose restrictions on parameters of production functions, so that they are able to approximate the stylised facts of economic behaviour that neoclassical economists generally agree characterise the real world (Chambers, 1988). On the production function, these restrictions regard:

    - monotonicity (and strict monotonicity) of the production function;
    - quasi-concavity (and concavity) of the production function;
    - weak essentiality (strict essentiality);
    - closed and non-empty input requirement set for all outputs;
    - the production function is finite, non-negative, real valued for all non-negative and finite inputs;
    - the production function is everywhere twice-continously differentiable.

    It has been underlined, however, that these restrictions have generally been rejected in econometric analyses (Evans and Heckman, 1984). The choice to impose these restrictions even if they have been rejected by empirical data, as is often the case, is not required by any direct comparison between our analysis and other studies where the restrictions have been imposed. Moreover, in our particular case the economic reasoning behind these analytical restrictions is difficult to accept a priori. Our production function is in fact a quasi-production function (where inputs are more than the conventional capital and labour inputs), representing an aggregate economic behaviour which may not immediately follow the same economic rules imposed by the neoclassical individual firm's behaviour.

6 Large Firm Acquisitions, Spin-Offs and Links in the Development of Regional Clusters of Technology-Intensive SMEs

ÅSA LINDHOLM DAHLSTRAND

6.1 Introduction

Large and small firms interact in many ways. Different buyer and seller relationships are important for the survival of all firms, and this is probably also the most usual kind of large firm-small firm link. However, especially among high-technology or technology-intensive SMEs, research collaboration and other RTD-related links with large firms ought also to be very important. This chapter focuses on technology-related ownership changes between large and small high-technology firms. These include the spin-off of new SMEs from large firms, intermediate links between large and small firms, and, finally, large firm acquisition of technology-intensive SMEs.

A regional concentration of an industry often acts as a strong magnet attracting both knowledgeable persons for future recruitment and external large, and perhaps multinational, firms. External large firms entering a region often do this by establishing a subsidiary, but they may also enter by acquiring the ongoing operations of a local firm. Within the TSER European network, it has been suggested that large local high-technology firms may be a positive and significant influence on the growth of regional SME clusters (Keeble and Wilkinson, 1998). Partly, this is because of the large firms' important role as a source of technology-intensive spin-offs. In addition, after such a spin-off firm is established it is usual that a continued link – especially for research collaboration – is established between the large firm and the new spin-off firm. With time, and the increased size of the new

Large Firm Acquisitions, Spin-Offs and Links 157

firm, these links are often complemented with new links outside the region. When such links include research collaboration with 'external' large firms, they may result in the large firm wanting to acquire the high-technology SME. If it does not lose its autonomy, this acquired high-technology SME might then be used not only for a small firm-large firm link, but also (by being local subsidiaries of large global firms) as a link between the regional milieu and the global business of the large firm.

The focus of this chapter is on the interaction between large and small technology-intensive European firms. Empirical evidence from research by members of the TSER European network will be used to consider the following questions:

- which high-technology regions and industries are dominated by large firms and SMEs, respectively;

- to what extent do large firms spin-off new local technology-intensive SMEs;

- do local large firms act as incubators, and are links between incubating firms and spin-offs more usual and/or more important than links with other firms;

- how common is it for large firms to acquire local technology-intensive SMEs;

- are such acquisitions primarily made by large firms internal or external to the region;

- can the large multinational firm act as a link between regional high-technology clusters and networks?

All regions cannot be assumed to behave similarly. In fact, different regions studied by the members of the TSER Network have quite different large firm-small firm interaction and linkage patterns. These differences may reflect the institutional framework, the industrial structure, or the technological or sectoral focus of a region. This chapter analyses patterns of large firm-small firm linkages, collaboration and ownership changes in the 10 European regions studied by the network. First, in the following section, some comments and considerations based on earlier findings will be made on the interaction between large and small innovative firms. Following on this, the 10 European regions will be discussed with respect to each area's industrial structure, sectoral focus and networking pattern. The following section examines the birth of new local technology-intensive firms in terms of the spin-off process. A further section discusses continuing relationships between parents and spin-off firms, as well as other links and co-operation between regional large and small technology-intensive firms. The final section presents preliminary conclusions on large firm acquisitions in these 10 European regions. Particular attention will be paid to the role of multinational large firms, since these may act as a link between different regional innovative networks or milieu.

6.2   Interaction between Large and Small Innovative Firms

A considerable amount of research has focused on the question of whether large *or* small firms are the most frequent, effective and/or efficient innovators, rather than on the interaction between these firms. However, large and small firms interact in several ways, for example, through different types of co-operation and technology-related ownership changes. While the large firm has innovatory advantages because of its relatively greater financial and technological resources, small firms instead have advantages of entrepreneurial dynamism, internal flexibility and responsiveness to change (Rothwell and Dodgson, 1994). In sectors with a high cost and long pay back period of research, large firms may have an especially important potential as sources of innovation. Usually, a large technology-based firm encompasses several technologies within its existing product areas, and this firm may have a potential advantage in a relatively larger amount of technology development. The technologies can also have a potential for innovation outside the existing product areas of the large firm. It is, however, usual that a large firm does not encourage potential innovations outside its existing product areas. Thus, the large firm may have an often neglected potential, as the source of radical new innovations. By spinning off this potential for external development, the large firm can act as an important source of innovation. Moreover, since new spin-off firms are most often set up geographically near their spin-off parents (e.g. Lindholm Dahlstrand, 1999), the spin-off process underlines how important already existing organisations are for the local establishment of new firms and for regional development. Because the two major sources of new high-technology small firm entrepreneurs are higher education institutions and well-established industrial firms (Oakey, 1995), new firms cluster around universities, research organisations and existing firms. This results in a natural tendency towards a substantial and probably growing disparity between regions that already possess indigenous high-technology activities, and those that do not (Keeble and Oakey, 1997).

Because of material advantages, e.g. in management, marketing, finance and R&D, large firms are able quickly to exploit inventions on a large scale. Williamson (1975) discussed the organisational limits of the large firm, including among other things the problems of bureaucratic commitment and entrepreneurial incentives. Thus, it seems clear that both the newly established entrepreneurial small firm and the large-sized firm have their respective advantages in the progress of technological development and change. This is also one important ingredient in Williamson's recommendation of a 'systems solution by classical specialization'. In brief, his argument is that small firms are frequently high performers when it comes to product innovation. Furthermore, small firms often have advantages in the early stages of the innovation process, as well as in less expensive and radical innovations, while large firms have an advantage in the later stages of scaling up innovations. In other words, large scale, or size, is often found to be a determinant of malfunctioning in the earlier and creative stages of the innovation process. Therefore, the 'systems solution by classical specialisation' may be an efficient innovation process. Williamson hypothesised that, because of the large firms' innovative disabilities in the early stages, an efficient procedure by which to introduce new products would be to allow the initial stages of the innovation process to be performed by independent innovators and small firms. The successful companies would then be acquired for subsequent development by a large firm. Large firms have been found frequently to diversify through a process of such small-firm acquisitions; one example here is how large Japanese firms have acquired both US and European small multimedia and biotechnology firms (Rothwell and Dodgson, 1994). Besides acquisitions, also different kinds of co-operation and alliances with firms based in other countries are becoming a popular way of gaining foreign market entry (McNally, 1996).

Technology-related spin-offs and acquisitions have been studied as part of an 'economic system of ownership changes' (Lindholm, 1994, 1996a). This system, as illustrated in Figure 6.1, links together small and large technology-intensive firms through different relations and ownership changes. The technology-related spin-offs in this system can mainly be of four different kinds: (1) a direct divestment; (2) the creation of an entrepreneurial new firm, subsequently acquired by another firm; or (3) continuing as independent; or (4) creation of a new firm that is later re-integrated into the spin-off parent organisation. An internal spin-off occurs when one part or unit of a company is transferred to another part of the same company. The internal spin-offs are included in the fourth category, since this category reduces to an internal spin-off when the time period as an independent firm is zero. In general, one of the main differences between the different spin-off categories is for how long the firm is acting independently.

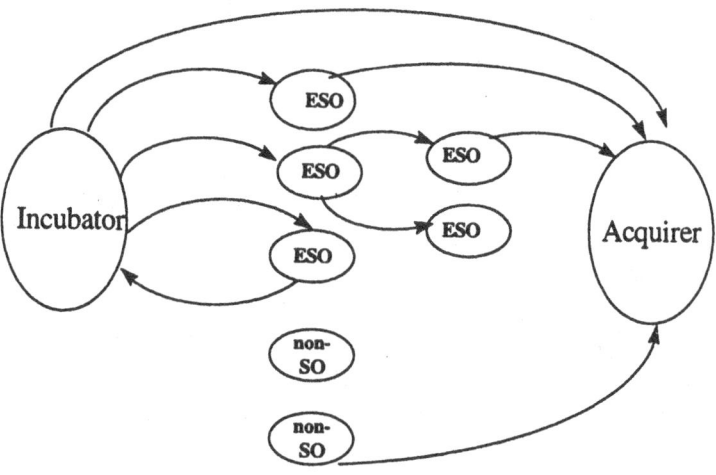

Figure 6.1 An economic system of ownership changes
Source: Lindholm, 1994, 1996a.

Building and maintaining the technology base can include internal development, as well as acquisitions and spin-offs of technology-based firms. Lindholm (1994) found that regardless of whether the initial technology was developed in an externally acquired unit or an internal development project, it is typical of technology-related spin-offs that some technology development is carried out in a private firm. In a system of firms interacting through ownership changes, the technology-related spin-off can be considered as a first mechanism, and acquisitions of technology-based firms as a second mechanism, within the system. Similarly, the acquisitions can be made as: (1) direct acquisition of other firms' divestments; (2) acquisition of other organisations' entrepreneurial spin-offs (ESOs); (3) acquisition of independent non-spin-off firms (non-SO); and (4) acquisition of an earlier spin-off from one's own organisation (internal spin-off). Furthermore, many of the ownership changes are made gradually, often initiated in different collaborative arrangements and links between firms, and as a result it is possible for a small firm to co-operate with and have several large firms as minority owners, for example both an earlier spin-off parent and a future owner.

As suggested by Rothwell and Dodgson: 'when the inter-company relationships generally are intensifying, and innovation is becoming even more of a networking process, the ability of firms of all sizes to forge mutually complementary relationships will be an increasingly important factor in the creation of competitive advantage through innovation' (1994, p. 321). In addition to the local spin-off process, many researchers have found that proximity is also important for the networking process and interaction in various ways, especially for small and medium sized enterprises or SMEs (Saxenian, 1994; Storper, 1993, 1995; Sternberg, 1996; Keeble, 1997). Besides the importance of proximity, the firms in a regional system are often assumed to benefit from shared – or similar – cultural and institutional frameworks. Furthermore, the globalisation of economies appears to be associated with a decline in the importance of traditional localised factors of production, whereas those parts of the economy that are based on knowledge and learning are subject to increasing returns (Malmberg, 1997). The positioning of specific regional sectoral clusters, for example based on local educational and technological strengths, can act as a prime locational pull-factor, affecting the spatial distribution of investments made by large multinational firms (see chapter 2, sections 2.2 and 2.7, and Ivarsson, 1997). In a globalised economy, inward foreign direct investments, including acquisitions, seem to be pulled to countries and regions that can offer the best complementary assets and quality of infrastructural support. Hence, many workers now argue that globalisation implies that location matters even more than in the past (Cantwell and Iammarino, 1998).

In a knowledge-based economy, regional networking, research and technology development, and collective learning may be important key processes for future local development and attractiveness. Lorenz (1996) points to four different mechanisms which may result in a base of common knowledge and shared rules which facilitates the development of a regional innovative milieu: (1) there might be common on-the-job training experience gained in larger firms or universities in the region; (2) the rules of the local labour market may influence the mobility of labour across SMEs in the region; (3) spin-offs from universities and large firms may have a role in establishing a base of shared knowledge; and (4) different technological alliances and user-producer relations might give rise to collective learning processes. Once again, the local nature of the spin-off process is underlined. Local large firms influence collective learning and technology development through their educational role. Large firms can act as important educators both when it comes to the education and training of technical and other key personnel, but also as incubation organisations for new spin-off firms. When such a spin-off is generated deliberately, the private industrial firm may also be the future partner in an alliance or a user-producer relation. Collective learning mechanisms can however be both 'conscious' and 'unconscious', and the transfer of creative and cumulative knowledge can take place even against the will of the first inventor (Capello, 1998), that is, for example, when a spin-off is made against the will of the parent organisation. Spin-offs may indeed be very important as a mechanism for unconscious collective learning within a region. They may even be more frequent and significant than conscious co-operation, for example through formal research collaboration.

The spin-off process is likely to enhance regional knowledge development and learning processes because it involves the diffusion and sharing of technological and managerial expertise within the region, promotes the creation of a common technology- and research-focused regional industrial culture, and encourages the development of inter-organisational links and personal networks through which new technology and knowledge can be shared and created. The externalisation disadvantages of the technology-related spin-offs may be of limited importance compared to the classical specialisation advantages of large and small technology-based firms. In the spin-off process these include several benefits from the disintegration, e.g. through removal of internal innovation barriers (in the source organisation). While Williamson recognises benefits of the systems approach, where the transaction costs are lowered because of the internalisation advantages when the small firm is acquired, he overlooks the potential benefits of classical specialisation where small firms are spun off in order to take advantage of the early-stage technological *innovation* advantages. In other words, regional spin-offs can act as carriers of knowledge and be important for an efficient innovation process. Although transaction-cost theory contributes to the understanding of why firms co-operate, it does not offer a complete explanation. Cost minimisation is not the only reason for firms to interact through different kinds of co-operation and ownership changes, and, thus, the importance of transaction costs must be compared to the importance of having technological innovations more fully exploited. Both transaction costs and technological change – especially when it comes to innovations – are interrelated and important for different organisational arrangements, and the benefits may alter over time in a dynamically changing pattern of organisational arrangements. Both inter-firm and inter-regional networks can provide a variety of such arrangements, allowing both large and small firms to interact, and also, the advantages of 'globalisation' and 'regionalisation' of innovative activities to be considered. Here the large multinational firm serves a key role; it can be an important local training and educational ground, but it can also, sometimes through its acquisitions, be considered as the key-ring of the chain from global to local (Cantwell and Iammarino, 1998).

6.3 Large and Small Firms in the 10 European Regions

In this section aspects of the industrial structure and technological focus of the 10 European regions studied by the members of the TSER Network will be discussed. A schematic picture of different regional characteristics may be painted by using Longhi and Keeble's (chapter 2) discussion of 'apparent' characteristics. In this, they identify four broad types of innovative regions.

1 *University-based regions* are local systems whose high-technology industrial dynamics have originated from the accumulated knowledge base of their universities. Keeble and Longhi (chapter 2) consider Cambridge, Oxford and Pisa to belong to this category.

2 *Industrial regions*. Keeble and Longhi (chapter 2) consider Europe's industrial regions (e.g. Munich, Grenoble and Göteborg) to have faced a deep crisis of restructuring during the last few decades. The regions have developed their collective capabilities allowing rapid adaptation to change, but many of the industrial regions are still in transition towards new forms of activities.

3 The *metropolitan area* is viewed by Longhi (1996) as being characterised by the domination of market relationships internally and co-operative relations between components of the local system and their external economic environment (e.g. Milan). Silicon Valley is a further example where the locality benefits greatly from its external relationships. The high rate of spin-off and sell-out is an important component here.

4 *Technopoles and peripheral regions.* Longhi (1996) considered peripheral areas to be categorised as externally-dominated systems. Local firms have their market- and co-operative relations outside the region, often based upon external investments. Examples are high-technology areas that have developed through the influence of public policy (e.g. Sophia-Antipolis).

Table 6.1 classifies the 10 regions in terms of these four categories. A brief review of each follows.

Cambridge

The number of Cambridge region high-technology firms increased from 30 in 1960 to 798 in 1998 (Keeble et al., 1999). Such firms accounted for around 20 per cent of total regional employment in 1996 (Keeble, 1998). The region is characterised by several dynamic micro-clusters of SMEs, with recent employment growth in high-technology services. Moreover, the great majority of the region's technology-based firms are small and locally founded, often as a result of a spin-off process. The Cambridge SME cluster can be considered as having evolved primarily through dynamic local processes of small firm start-up and spin-off, either from the university or from local SMEs. Many of the larger technology-based firms have been acquired by external large multinational firms. Several large multinational firms have recently set up laboratories in the region in order to tap into local technological and scientific research competencies (Keeble et al., 1998).

Oxford

The Oxford region exhibits some indications of an emerging 'innovative milieu', but there are still a number of constraints which inhibit the development of local networks and collective learning (Lawton Smith, 1998). In the 1960s, Oxford was both a university town and a centre for car manufacturing. Since then the region has been transformed by the growth of its universities and high-technology industries. Reflecting the region's industrial tradition, the majority of firms are small manufacturing enterprises, serving specialist niche markets. But there are also a number of large firms, both locally-headquartered and multinational. Several multinational firms have set up R&D laboratories in the region. The region has a developing concentration of firms in software, biosciences and motor racing. Many bioscience firms are spin-offs from Oxford University.

Italy

Studies of the 'Third Italy' demonstrate a need to depart from the traditional tripartite Italian model. A clear-cut distinction emerges between the north-west and north-east of the region on the one hand, and central areas on the other. The latter have only been able to maintain a good competitive performance through severe employment cuts (Capello and Camagni, 1998). The northern high-technology districts (Piacenza, Pisa and NE Milano) analysed here have achieved significant industrial employment growth. While Pisa may be classified as a university based region, Milan can instead be considered a metropolitan area without important, stable and remarkable linkages between universities and SMEs. Spin-offs are a distinctive feature of the Italian regions, comprising 89 per cent of the high-technology SMEs studied (72 per cent from industrial firms and 17 per cent from universities). The role of large firms has not been analysed by the TSER team, and their importance cannot therefore be measured. Nevertheless, Capello and Camagni (1998) find that collective learning is more linked with small firms and with radical innovation processes.

Grenoble

Grenoble is a region that is characterised by both spin-offs and acquisitions. It can be classified as an area in transition, with a successful industrial history. In addition to several large firms, it contained 11,000 micro firms in 1992, and an additional 1,700 larger SMEs which had survived between 1975 and 1992 (de Bernardy, 1998). Together these SMEs generated over 31,000 jobs 1975–92, a period during which the large firms fired over 13,000 employees. There are many examples of local small firm spin-offs from large multinational companies, such as Hewlett Packard, Bull and Télémécanique. However, there have also been many large firm acquisitions of small high-technology firms. Despite losing their independence, these have not usually moved away from the region after the acquisition. De Bernardy (1997) distinguishes two different trajectories of high-technology SME development in Grenoble; one half of the firms have been acquired and integrated into larger firms in the area, the other half of the SMEs have resisted and stayed independent, instead co-operating with other local and European firms.

Munich

Munich is Germany's leading high-technology region (Sternberg and Tamásy, 1999). It is dominated by a few large firms, but the great importance of such companies in Munich's high-technology sector offers potential opportunities for local SMEs (Sternberg and Tamásy, 1998). For example, Siemens plays a central role in the development of local high-technology SMEs. No less than 78 per cent of such businesses have Siemens as a customer. Moreover, 6.5 per cent of all SMEs surveyed by Sternberg and Tamásy (1999) were spin-offs from this single firm. This is comparable to spin-offs from local research institutes, which appear to be under-utilised in the region; approximately 6 per cent of the surveyed SMEs are spin-offs from the dominating Ludwig Maximilian University. Sternberg and Tamásy (1998, 1999) conclude that there is a considerable volume of intraregional learning between large and small firms in the Munich region.

Randstad/Utrecht

In his research on high-technology SMEs in the Randstad, Wever (1998) argues that regional differences in the Netherlands are very limited, and that the country should be seen as one 'urban field'. The Randstad region should thus primarily be considered as part of this field rather than as a specific regional innovative milieu. Nevertheless, Wever (1998) shows that in absolute terms, the Utrecht region contains the largest number (1732) of high-technology firms in the Netherlands, over 75 per cent of which belong to the IT sector. Forty percent of the high-technology SMEs which he surveyed were created as spin-offs from other firms. None (!) of these spin-offs had its roots in a university; 59 per cent had large firms as parents and 41 per cent had small firms. The majority of these spin-offs still co-operated in some way with their parent firm, and most spin-off founders regarded their parent firm as very important for the development of their innovative activities. Wever (1997) reports that exactly 50 per cent of these large spin-off parents were located in the region, while smaller incubating firms were even more often located inside the region. Links with customers and suppliers were important for innovative activities, but the majority of these links were with firms outside the region (76 per cent of customer links and 47 per cent of supplier links).

Göteborg

The Göteborg region possesses a long industrial history. It is dominated by several large manufacturing firms, primarily in the automotive industry. Besides large firms, the region also contains numerous small technology-intensive SMEs, especially within the mechanics and software sectors. Many of these SMEs have spun off from local industrial firms, but there are also many university spin-offs in the region. Approximately 45 per cent of surveyed Göteborg technology-based SMEs have founders who were previously employed in large industrial firms, but there are also many new firms created as second generation spin-offs from SMEs. Most spin-offs report close links with their parents, but over time these are complemented by external Swedish and international links. Technology-intensive Swedish SMEs are frequently acquired (Lindholm 1994, 1996b), but among the Göteborg SMEs only a sixth of the surveyed firms were acquired (Lindholm Dahlstrand, 1999). Most were acquired by large multinational firms from outside the region.

Helsinki

Helsinki is much the largest city of the county of Uusimaa, and clearly a metropolitan region in terms of Longhi's classification. Approximately 95 per cent of all industrial companies in the county are SMEs. In 1995 the region contained 5,300 industrial manufacturing firms, with paper, metals, machinery and electrical equipment the leading sectors. Partly due to subcontracting from Nokia and Ericsson, a high proportion of SMEs manufacture electrical equipment (Kauranen and Mäkelä, 1998). If an inventor in Finland (or Sweden) is employed by a university, this individual, rather than the university or the funding organisation, owns the intellectual property rights him-/herself. This encourages university-based inventors to develop spin-off companies, a process also supported by the government-funded Spinno programme. As a result, the region has experienced a large number of such spin-offs, with 170 supported by the programme since 1991.

Barcelona

Catalonia, focused on the metropolitan area of Barcelona, is a region in transition, and perhaps Spain's most economically important region. It contains many large multinational firms, the seven largest firms all being foreign owned. Escorsa, Maspons and Valls (1997) point out that these multinationals have been important for the region's economic development, for example through creating new jobs, but also through transferring technology to subcontractors and opening up new export markets. However, in the 1990s dependency has become excessive, with the consequent loss of decision-making capacity. Catalonia is not a region with a high level of R&D activities; many of the large multinationals carry out research in the parent company and not in the region (Escorsa, Maspons and Valls, 1998). The Barcelona region also provides examples of foreign firms acquiring local firms and then closing down the local R&D department. There are eight public universities in the region, but the number of research agreements between SMEs and these universities is low.

Sophia-Antipolis

Sophia-Antipolis was created by government policy, and the attraction of branch units of large national and multinational firms. Their market relations were initially European-wide, and they possessed very few local innovative links. However, in the 1990s (Longhi, 1997), this situation began to change; large international corporations were forced to downsize their operations and local spin-off SMEs in the information technology sector became much more important for Sophia-Antipolis' continued growth. Sophia-Antipolis is no longer exclusively dependent on large firms.

From Table 6.1 it is clear that there are important differences between the innovative regions analysed by the TSER Network. The four categories described by Keeble and Longhi (chapter 2) are all represented, and quite evenly distributed, in the table. There are large firms in all regions, but in a few regions these are primarily large external multinationals. Several regions, such as Cambridge, Grenoble and Sophia-Antipolis, have deliberately attracted investment by such firms, in the form of local establishments and subsidiaries. In Cambridge, large firm subsidiaries in fact exhibit a higher number of all kinds of inter-firm links (Lawson, 1997).

In addition, such multinational firms are also now encouraging regional small firm start-ups through provision of venture capital funding (Keeble et al, 1998). This is also encouraged by Cambridge University. In areas where a negative attitude may exist towards large multinational firm acquisition of local SMEs, providing venture capital may be superior since it can allow a 'window-on-technology' without causing trouble because of 'loss of independence' in the SME.

In order to analyse which industries and technological fields are dominated by large firms or SMEs, Table 6.2 summarises the sectoral specialisation of the different regions.

From Table 6.2, it is clear that high-technology services are usually dominated by SMEs. This sector is represented in all the European regions studied. A very different pattern is found in the chemicals sector, where SMEs are rare. The three other sectors, electronics, mechanical engineering and biotechnology/medicine, are usually characterised by a mix of large and small firms. In turn, this may be connected with the importance of large firms as a training ground in these sectors. Large firms can be both substantial recruiters of personnel and providers of trained personnel, and as such have an important educative function for small firms in the region. It may also be that SMEs operating in the same industrial sector can benefit from large firms attracting suppliers of specialist services into the area. However, one of the most educative roles fulfilled by the large firm is in supplying entrepreneurs for the corporate spin-off process.

Table 6.1 Large firms and categories of innovative regions

	Large firm		Innovative region			
	Local	Multi-national	University based	Industrial region	Metropolitan area	Peripheral/ transition area
Cambridge		X	X			
Oxford	X	X	X			
Italy		(X)	X		(X)	
Grenoble	X	X		X		(X)
Munich	X	X		X		
Randstad/Utrecht		X			X	
Göteborg	X	X		X		
Helsinki	X	X			X	
Barcelona		X			(X)	X
Sophia-Antipolis		X				X

Table 6.2 Sectoral and technological specialisation

	Biotech, medicine	Manufacturing			Service
		Electronics	Mechanics	Chemicals	IT/software/ etc.
Cambridge	SME	SME/L		(SME)	SME/L
Oxford	SME/L	L	L		SME
Italy			SME		SME
Grenoble		L/SME			SME
Munich	SME	L/SME	L/SME		n.a.
Randstad/Utrecht	L	L/SME		L	L/SME
Göteborg	L/SME	L	L/SME	L	SME
Helsinki		L/SME	L	L	SME
Barcelona	L/SME	L	L/SME		SME
Sophia-Antipolis	L/SME	L		L/SME	SME

Note: L = Large firm.

6.4 Corporate Spin-Offs

The TSER Network's research provides powerful evidence of the importance of entrepreneurial spin-offs, often from large firms, in the growth of technology-based industries and regions. The concept of 'entrepreneurial spin-offs' (ESOs) was originally proposed by Lindholm (1994). These spin-offs occur when an entrepreneur leaves an existing job to start a new firm, taking with him or her a product idea which originated in the previous firm or organisation. This must also include the transfer of some rights – assets or knowledge – from the existing legal body to the new firm. ESOs themselves can be categorised depending on what legal body, or organisation, they are spun off from, and on where the entrepreneur has gained his/her background experience. Lindholm (1994) subdivided ESOs into (1) university spin-offs (USOs), (2) corporate spin-offs (CSOs), or (3) other 'institutional spin-offs'. Entrepreneurs can also have several sources of background experience and there may be an 'incubator system'. In this chapter, however, the ESO concept is subdivided with respect to the latest employer, that is, the 'main incubator' or 'spin-off parent'. Table 6.3 summarises the importance of both universities and large and small firms as spin-off parents in the 10 European regions.

Table 6.3   Entrepreneurial spin-offs

	LCSO	SCSO	USO
Cambridge	–	X	X
Oxford	X	X	X
Italy	–	X	X
Grenoble	X	–	X
Munich	X	–	X
Randstad/Utrecht	X	X	–
Göteborg	X	X	X
Helsinki	(X)	n.a.	X
Barcelona	n.a.	n.a.	n.a.
Sophia-Antipolis	(X)	n.a.	n.a.

Notes

LCSO – Large firm corporate spin-off.
SCSO – SME corporate spin-off.
USO – University spin-off.

Table 6.3 shows that most regions have various sources of spin-off entrepreneurs. The Randstad/Utrecht area is the only region that reports no university spin-offs. It also reports numerous spin-offs from large multinational firms. Interestingly, Wever (1997) shows that spin-offs in the Utrecht region are larger (more employees) than other local SMEs. This suggests, in line with Lindholm Dahlstrand's (1997) finding for Göteborg SMEs, that corporate spin-offs grow faster than other SMEs, including university spin-offs. Even within one region, different large firms can of course have different spin-off strategies. Lorenz (1996) has illustrated this with an example from Minnesota: 3M encourages spin-offs, whereas Metronics aggressively tries to prevent spin-offs.

Large firms are also an important source of high-technology spin-off entrepreneurs in Oxford, Grenoble, Munich and Göteborg. In the Göteborg case (Lindholm Dahlstrand, 1999), they appear to be extremely important in training entrepreneurs for both spin-offs and non-spin-offs. This contrasts with the prevailing wisdom on sources of new firm founders generally, which is that these are generated primarily from small firms which provide opportunities and role models for gaining experience of management and risk-taking (Dorfman, 1983; Storey, 1982, 1994). In university-based regions (Cambridge, Oxford and Italy/Pisa), the majority of CSO founders do come from small firms. In turn, this further illustrates the importance of the region's existing industrial structure.

It might be hypothesised that Finnish and Swedish intellectual property rights of university employees could lead to an exceptionally high frequency of USOs in these countries. This is not, however, supported by the findings of the TSER network; Table 6.3 indicates that university spin-offs are important in all but one region. A special category of local firms, leading R&D consultancies, with strong links to Cambridge university, has played an important role as a breeding ground for new spin-off firms in the Cambridge region. These consultancies have often encouraged employees to set up new firms, stimulating spin-off activity and the evolution of a regional business culture characterised by collaboration and co-operation. In the Oxford region, Lawton Smith (1997) points to the great importance of Oxford Instruments. Approximately 20 firms have been spun off directly from this local large firm, stimulating further activity and second generation spin-offs. In several of the regions, 'second generation' spin-offs are frequent. In both Oxford and Cambridge, nearly half the SMEs surveyed have spun-off additional new SMEs of their own. In the Göteborg region, 52 technology-intensive SMEs have spun off an additional 26 new firms (Lindholm Dahlstrand, 1999).

6.5 Links between Large Firms and SMEs

This section examines different links and networking between large and small firms, and what forms of co-operation characterise such links in the different regions. Do incubating firms and spin-offs continue to co-operate over time or do links with other firms become more important? Large firms may be important as both buyers and sellers of SME products and services. In the most advanced case, large firms can engage in 'sponsored spin-offs' (Lindholm, 1994; Rothwell and Dodgson, 1994) where they may continue to own a minority share of a spin-off firm and offer financial contributions for the development of the firm. The sponsored spin-offs can be designed to include several other kinds of assistance and links between the parent and the new firm. Moreover, links and co-operation can be both local and external, while large firm links can help the internationalisation process of an SME. In Table 6.4, the most important links for co-operation between large and small firms are exemplified for the different regions. Important links with the spin-off parent or an acquiring parent organisation are also included in the table.

Table 6.4 Local and external links between SMEs and parents or other large firms

	Parent – SME link	Local large firm – SME links	External large firm – SME links
Cambridge	spin-off	n.a.	n.a.
Oxford		suppliers and subcontractors	n.a.
Italy	spin-off	n.a.	n.a.
Grenoble	acquisition	customer, subcontractor and RTD	customer, subcontractor and RTD
Munich	(spin-off)	large firms are customers, 1/3 of SMEs RTD links to Siemens	n.a.
Randstad/Utrecht	spin-off	a few supplier links	customer links
Göteborg	spin-off	customer links	customer and supplier links
Helsinki	n.a.	n.a.	n.a.
Barcelona	acquisition	a few customer links (sectoral variation)	RTD-links
Sophia-Antipolis	n.a.	with large firms in the park	n.a.

At least half of the regions report a continued relationship between local spin-offs and their parent firm. In many cases such links are the basis for a continuing local customer or supplier relationship. These links are also the most frequent local links between technology-intensive SMEs and large firms. Links for RTD-co-operation are unusual in Table 6.4, but this might be explained by the fact that many technology-intensive SMEs consider research as the 'product or service' they are selling to their large firm customer. Unfortunately, the data in Table 6.4 does not allow any conclusions to be drawn about other links between SMEs and external large firms. However, in both Grenoble and Barcelona there are many important links to acquiring organisations. In Grenoble these are often local links, but in Barcelona the many foreign multinationals have strong subsidiary-parent links with external regions.

In the Munich region, Sternberg and Tamásy (1999) found the interaction between Siemens and local high-technology SMEs to be beneficial to both the large and the small firms. On the one hand, Siemens was found to co-operate with local technology-intensive SMEs since the large firm can profit from having such a 'window-on-technology'. On the other hand, Sternberg and Tamásy (1999) consider Siemens to be strengthening local innovativeness by affording local SMEs access to their much larger (international) network. Unfortunately, the study does not reveal whether or not this leads to Siemens acquiring any of these small firms.

In both the Göteborg and the Utrecht areas, collaboration and co-operation between spin-offs and their parents appears to be the most important local link. In the Dutch case 55 per cent of spin-offs still had a continuing co-operative relation with their parent. In the Göteborg region the corresponding figure is 60 per cent of the corporate spin-offs. For other SMEs (including other spin-off categories), the links to the former employer became much less important over time.

It may be argued that proximity matters for the genesis and intensity of innovative linkages between large firms and SMEs. But its importance varies between the 10 regions analysed here. It is obvious from the above that there are different links between large companies and SMEs in different regions. In turn, this could partly be explained by the large firm's interest in collaborating with SMEs because of their considerable competencies and specialisation in narrow niches. Large firms may find it more economic to collaborate than to develop their own expertise and may wish to acquire the SME in time. The innovative capacity of technology-intensive SMEs is a key factor in this.

6.6 Large Firm Acquisitions

In Sweden most successful small technology-based firms are in due course acquired by large firms in order to access their technological expertise. New technology often involves high levels of tacit knowledge, and it may be essential that human capital in an effective team configuration accompanies the transfer of technology, i.e. 'embodied knowledge'. This is also one of the greatest advantages when technology is transferred through spin-off or acquisition of a technology-intensive company. Acquiring a whole technology-intensive firm (or even partially, via formalised RTD-co-operation, like a joint venture), rather than just a license for its technology, is then a way of making a package deal, which includes the typically non-tradable complementarities of the SME.

In turn, SMEs often want to be acquired, usually because they need capital for further development or for internationalisation (Lindholm 1994, 1996b; Karaömerlioglu and Lindholm Dahlstrand, 1999). Swedish evidence further suggests that technology-based SMEs grow faster after acquisition, and retain their innovativeness. This may be because in Sweden, large firms leave acquired SMEs with their existing autonomy. Garnsey (1996) has suggested that the negative side of the acquisition process should be balanced against the idea that such acquisitions are an inevitable part of the life cycle in many cases.

Large external acquirers can be important in the development of innovative regions. For one thing, they may bring with them new knowledge that can be exploited within the area (Cantwell and Iammarino, 1998). The large multinational may also be attracted by a region since it wants to tap into local innovative capabilities. For example, several large Swedish multinationals have been attracted to Northern Italian regions (Berggren et. al., 1998). For a large multinational firm, an acquisition may be more attractive than setting up a new subsidiary in a region where it wants to be established, particularly since regions experiencing rapid technological development usually contain considerable tacit knowledge embodied in local firms and individuals. This may be a reason why, as mentioned earlier, Japanese firms have been found to acquire both US and European small multimedia and biotechnology firms (Rothwell and Dodgson, 1994).

Data on large firm acquisitions is unfortunately not available in all 10 European regions. A few examples are given in Table 6.5.

Table 6.5 Primary large firm acquirers

	Local	**National**	**Multinational**
Cambridge	No	No	Yes
Oxford	n.a.	n.a.	n.a.
Italy	n.a.	n.a.	n.a.
Grenoble	No	Yes	Yes
Munich	n.a.	n.a.	n.a.
Randstad/Utrecht	n.a.	n.a.	n.a.
Göteborg	No	(Yes)	Yes
Helsinki	n.a.	n.a.	n.a.
Barcelona	No	No	Yes
Sophia-Antipolis	No	Yes	No

One very interesting feature of Table 6.5 is that no region reports primarily having local acquirers. Of course, some of the regions lack large firms, other than external multinationals that have set up or acquired local subsidiaries. Acquisitions in Table 6.5 are in fact dominated by large multinationals acquiring local technology-intensive SMEs. The available data on acquisitions by national large firms are dominated by French examples. Here, Longhi (1996) suggests that acquisition activity in Sophia-Antipolis, which has developed during the 1990s, has had positive effects. Existing large firms such as France Telecom had been overdependent on old technologies, and new technology-intensive SMEs have provided these firms with access to new innovative technologies with which they could restructure their activities.

Research in the Barcelona region (Escorsa, Maspons and Valls, 1998) suggests high rates of acquisition of local companies by foreign multinational firms, with almost half of the 'decisive' companies in each industrial sector now in foreign hands. This could have serious negative effects if the tendency to close down local RTD-departments of acquired firms persists. Tapping into and contributing to regional innovativeness seems to be more usual in, for example, the Cambridge case. Here, many successful technology-intensive SMEs have been targets for acquisition by large multinational companies. For example, in the industrial and scientific instrumentation sector, Garnsey (1998) found that eight out of the 14 leading local firms had been acquired by large firms external to the region. Moreover, as noted by Keeble et al. (1998), many current local start-ups do appear to welcome the prospect of being acquired in the future. Overall, however, only 10 per cent of Cambridge technology-based SMEs surveyed had been acquired. In the Göteborg region, the corresponding figure is one-sixth (Lindholm Dahlstrand, 1999), although nationally, most technology-intensive Swedish SMEs are eventually acquired (Lindholm 1994, 1996b). Göteborg SMEs are most often acquired by large multinational firms from outside the region. In another Swedish study, Ivarsson (1997) found that foreign direct investments in Sweden were usually made through acquisition in some of the nation's most competitive and successful industrial clusters.

6.7 The Large Multinational Firm as a Link between Regional Networks? Discussion and Implications

One of the basic questions initially posed by the TSER network was: 'Are successful clusters being mainly driven by endogenous processes of SME development, or can exogenous processes, perhaps associated with large firms or policy initiatives, also be successful?' (Keeble, 1995, p. 2). This is a difficult question to answer, but this chapter has provided some valuable evidence. Specifically, it has been argued that, because both acquisitions and spin-offs involve the transfer of tacit knowledge and embodied expertise, such ownership changes can promote both intra- and inter-regional links between firms, and thus play an important role for collective learning.

The TSER network's research provides convincing evidence of the importance of local spin-off processes in cluster evolution, and demonstrates that large firms are particularly important as spin-off parents in certain regions. Well known American examples like Silicon Valley and Minneapolis (Lorenz, 1996) are complemented by European cases such as Utrecht, Munich, Grenoble and Göteborg. In all these clusters large firms have played a very important role in the generation and development of technology-intensive SMEs. This is illustrated by the Utrecht case; Wever (1997) found that 82 per cent of SME founders considered the skills they acquired in their previous firm as very important for their SME's innovative activities.

Lorenz (1997) argues that large firms, through their ability to develop international links and their large number of partnerships, can act as a conduit for technology transfer and expertise into as well as out of the region. This is similar to the 'key-ring' role of multinationals, as key links in the chain from the global to the local (Cantwell and Iammarino, 1998). In Grenoble and Sophia-Antipolis, for example, spin-offs from multinational firms sometimes take with them technology acquired outside the region, thus channelling technology into the region. Moreover, the establishment of new local subsidiaries by large multinational companies has also been found to have positive and significant influences on the growth of regional clusters. Since as Longhi and Keeble stress (chapter 2), innovation often requires access to locational specificities, large firms are frequently very active in establishing local links. As noted earlier, the Cambridge survey (Lawson, 1997) found no evidence to suggest that subsidiaries are less embedded in the local system of inter-firm relations. Indeed, the evidence actually indicates the opposite, that local subsidiaries value certain local links, especially research links, more highly than independent firms. In turn, multinational innovativeness often benefits from linking local innovation networks to similar networks in other regions in which it operates. In this way, the large multinational firm may play a special role in transferring knowledge and skills between different regions, as part of a globalisation process.

The research reviewed above shows that multinational acquisitions are quite common in several of the European regions studied. Such acquisitions can have a negative impact if, for example, large firms acquire SMEs which might be future competitors. However, the evidence from Cambridge (Keeble, et. al., 1998), and also earlier Swedish results (Lindholm 1994, 1996b), does not on balance support such a negative judgement. On the contrary, large firm acquisition of local technology-intensive SMEs seems often to bring new resources and help with marketing and internationalisation. As argued by the 'system of ownership changes' approach (Lindholm, 1994, 1996a), large firm inflexibility often means that early-stage innovation is best performed within a new, small, spin-off firm. Later, however, with growth and the need for additional resources (capital, access to international markets), such firms can benefit from acquisition by another large firm, in a way that promotes innovation by utilising the advantages of both small and large firms.

The TSER network's research thus suggests that large firms are generally a positive and significant influence on the development of European regional clusters of high-technology SMEs. In some regions, this is because of the large firm's important role as a source of technology-intensive spin-offs. Such spin-offs also usually maintain beneficial links, especially research links, with their former parent. Over time, and with the growth of the new firm, these links are often complemented with new links outside the region. When such links include research collaboration with 'external' large firms, they may result in large firm – and usually multinational – acquisition of the high-technology SME. If an innovative local SME does not lose its autonomy after such an acquisition, it may well play a key role not just in local small firm-large firm networks, but also as a link between the regional milieu and the global activities of its parent multinational.

References

Berggren, C., Brulin, B. and Gustafsson, L.-L. (1998), 'Från Italien till Gnosjö', *RALF*, Rapport No. 2 (in Swedish).
Cantwell, J. and Iammarino, S. (1998), 'MNCs, Technological Innovation and Regional Systems in the EU: Some Evidence in the Italian Case', *Discussion Papers* in

International Investment and Management, Series B, Vol. X, Department of Economics, University of Reading, UK.

Capello, R. (1998), 'Collective Learning and the Spatial Transfer of Knowledge: Innovation Processes in Italian High-Tech Milieux', in D. Keeble and C. Lawson (eds), *Collective Learning Processes and Knowledge Development in the Evolution of Regional Clusters of High-Technology SMEs in Europe*, ESRC Centre for Business Research, University of Cambridge, Cambridge, pp. 19–38.

Capello, R. and Camagni, R. (1998), 'Regional Report: Italy', in D. Keeble and C. Lawson (eds), *Regional Reports*, ESRC Centre for Business Research, University of Cambridge, Cambridge.

de Bernardy, M. (1997), 'Small Firm – Large Firm Interactions', in D. Keeble and C. Lawson (eds), *Networks, Links and Large Firm Impacts on the Evolution of Regional Clusters of High-Technology SMEs in Europe*, ESRC Centre for Business Research, University of Cambridge, Cambridge, pp. 129–130.

de Bernardy, M. (1998), 'RTD SMEs and Collective Learning: Historicity and Ability for Local Economy to Evolve: The Grenoble Case Study', in D. Keeble and C. Lawson (eds), *Regional Reports*, ESRC Centre for Business Research, University of Cambridge, Cambridge.

Dorfman, N.S. (1983), 'Route 128: The Development of a Regional High Technology Economy', *Research Policy*, vol. 12, pp. 299–316.

Escorsa, P., Maspons, R. and Valls, J. (1997), 'Networks in a Big Metropolitan Area: The Case of Barcelona', in D. Keeble and C. Lawson (eds), *Networks, Links and Large Firm Impacts on the Evolution of Regional Clusters of High-Technology SMEs in Europe*, ESRC Centre for Business Research, University of Cambridge, Cambridge, pp. 3–24.

Escorsa, P., Maspons, R. and Valls, J. (1998), 'Regional Report: Barcelona', in D. Keeble and C. Lawson (eds), *Regional Reports*, ESRC Centre for Business Research, University of Cambridge, Cambridge.

Garnsey, E. (1996), in D. Keeble and C. Lawson (eds), *Report on Presentations and Discussions, Utrecht Network Meeting of the European Research Network on 'Networks, Collective Learning and RTD in Regionally-Clustered High-Technology Small and Medium-Sized Enterprises'*, ESRC Centre for Business Research, University of Cambridge, Cambridge.

Garnsey, E. (1998), 'Dynamics of the Innovative Milieu: Examples from Cambridge', (Summary by D. Keeble) in D. Keeble and C. Lawson (eds), *Collective Learning Processes and Knowledge Development in the Evolution of Regional Clusters of High-Technology SMEs in Europe*, ESRC Centre for Business Research, University of Cambridge, Cambridge, pp. 117–22.

Ivarsson, I. (1997), 'Generating Technology through Inward FDI in Competitive Host-Country Clusters', paper presented at the 1997 Residential Conference of the *International Geographical Union Commission on the Organisation of Industrial Space*, 3–9 August, Göteborg, Sweden.

Karaömerlioglu, D. and Lindholm Dahlstrand, Å. (1999), 'The Dynamics of Innovation Financing in Sweden', paper presented at the 44th ICSB World Conference, Naples, 20–23 June.

Kauranen, I and Mäkelä, J. (1998), 'The Final Report from Finland', in D. Keeble and C. Lawson (eds), *Regional Reports*, ESRC Centre for Business Research, University of Cambridge, Cambridge.

Keeble, D. (1995), Annex 1: Thematic Network on Networks, Collective Learning and RTD in Regionally-Clustered High-Technology Small and Medium Sized Enterprises, Research Proposal to the DG XII; the EU Commission.

Keeble, D. (1997), 'Small Firms, Innovation and Regional Development in Britain in the 1990s', in D. Keeble and C. Lawson (eds), *Regional Reports*, vol. 31, pp. 281–93.

Keeble, D. (1998), 'Local Industrial Development and Dynamics: The East Anglian Case', *ESRC Centre for Business Research, University of Cambridge, Cambridge, Working Paper* no. 96.

Keeble, D., Lawson, C., Moore, B., and Wilkinson, F. (1998), 'Regional Report: Cambridge', in D. Keeble and C. Lawson (eds), *Regional Reports*, ESRC Centre for Business Research, University of Cambridge, Cambridge.

Keeble, D., Lawson, C., Moore, B. and Wilkinson, F. (1999), 'Collective Learning Processes, Networking and "Institutional Thickness" in the Cambridge Region', *Regional Studies*, vol. 33, pp. 319–32.

Keeble, D. and Oakey, R. (1997), 'Spatial Variations in Innovation in High-technology Small and Medium-sized Enterprises: A Review', unpublished working paper, ESRC Centre for Business Research, University of Cambridge, Cambridge.

Keeble, D. and Wilkinson, F. (1998), 'Twelve-monthly Progress Report, February 1997–January 1998', for the TSER European Network on Networks, Collective Learning and RTD in Regionally-Clustered High-Technology Small and Medium-Sized Enterprises, ESRC Centre for Business Research, University of Cambridge, Cambridge.

Lawson, C. (1997), 'Local Inter-Firm Networking by High-Technology Firms in the Cambridge Region', in D. Keeble and C. Lawson (eds), *Networks, Links and Large Firm Impacts on the Evolution of Regional Clusters of High-Technology SMEs in Europe*, ESRC Centre for Business Research, University of Cambridge, Cambridge, pp. 41–56.

Lawton Smith, H. (1997), 'Inter-Firm Networks in Oxfordshire', in D. Keeble and C. Lawson (eds), *Networks, Links and Large Firm Impacts on the Evolution of Regional Clusters of High-Technology SMEs in Europe*, ESRC Centre for Business Research, University of Cambridge, Cambridge, pp. 57–66.

Lawton Smith, H. (1998), 'Regional Report: Oxfordshire', in D. Keeble and C. Lawson (eds), *Regional Reports*, ESRC Centre for Business Research, University of Cambridge, Cambridge.

Lindholm, Å. (1994), *The Economics of Technology-Related Ownership Changes: A Study of Innovativeness and Growth through Acquisitions and Spin-offs*, Department of Industrial Management and Economics, Chalmers University of Technology, Sweden.

Lindholm, Å. (1996a), 'An Economic System of Technology-related Acquisitions and Spin-offs', *ESRC Centre for Business Research, University of Cambridge, Cambridge, Working Paper* no. 33.

Lindholm, Å. (1996b) 'Acquisition and Growth of Technology-Based Firms', *ESRC Centre for Business Research, University of Cambridge, Cambridge, Working Paper* no. 47.

Lindholm Dahlstrand, Å. (1997), 'Entrepreneurial Spin-off Enterprises in Göteborg, Sweden', *European Planning Studies*, vol. 5, pp. 661–75.

Lindholm Dahlstrand, Å. (1999), 'Technology-based SMEs in the Göteborg Region: Their Origin and Interaction with Universities and Large Firms', *Regional Studies*, vol. 33, pp. 379–89.

Longhi, C. (1996), 'Large Firm-SME Linkages and Impacts', in D. Keeble and C. Lawson (eds), *Report on Presentations and Discussions, Utrecht Network Meeting of the European Research Network on 'Networks, Collective Learning and RTD in Regionally-*

Clustered High-Technology Small and Medium-Sized Enterprises', ESRC Centre for Business Research, University of Cambridge, Cambridge.

Longhi, C. (1997), 'Sophia-Antipolis: Recent Quantitative and Qualitative Trends', in D. Keeble and C. Lawson (eds), *Networks, Links and Large Firm Impacts on the Evolution of Regional Clusters of High-Technology SMEs in Europe*, ESRC Centre for Business Research, University of Cambridge, Cambridge, pp. 117–28.

Lorenz, E. (1996), 'Collective Learning Processes and the Regional Labour Market', in D. Keeble and C. Lawson (eds), *Report on Presentations and Discussions, Utrecht Network Meeting of the European Research Network on 'Networks, Collective Learning and RTD in Regionally-Clustered High-Technology Small and Medium-Sized Enterprises'*, ESRC Centre for Business Research, University of Cambridge, Cambridge.

Lorenz, E. (1997), 'Large Firm-Small Firm Interaction and Impacts in High-Technology Milieux', in D. Keeble and C. Lawson (eds), *Networks, Links and Large Firm Impacts on the Evolution of Regional Clusters of High-Technology SMEs in Europe*, ESRC Centre for Business Research, University of Cambridge, Cambridge, pp. 131–2.

Malmberg, A. (1997), 'Industrial Geography: Location and Learning', *Progress in Human Geography*, vol. 21, pp. 573–82.

McNally, K. (1996), *Corporate Venture Capital: Bridging the Equity Gap in the Small Business Sector*, Routledge, London.

Oakey, R. (1995), *High-Technology New Firms: Variable Barriers to Growth*, Paul Chapman Publishing Ltd., London.

Rothwell, R. and Dodgson, M. (1994), 'Innovation and Size of Firm', in M. Dodgson and R. Rothwell (eds), *The Handbook of Industrial Innovation*, Edward Elgar, Cheltenham.

Saxenian, A. (1994), *Regional Advantage: Culture and Competition in Silicon Valley and Route 128*, Harvard University Press, Cambridge, Mass.

Sternberg, R. (1996), 'Technology Policies and Growth of Regions: Evidence from Four Countries', *Small Business Economics*, vol. 8, pp. 75–86.

Sternberg, R. and Tamásy, C. (1998), 'Regional Report: Munich', in D. Keeble and C. Lawson (eds), *Regional Reports*, ESRC Centre for Business Research, University of Cambridge, Cambridge.

Sternberg, R. and Tamásy, C. (1999), 'Munich as Germany's No. 1 High-Technology Region: Empirical Evidence, Theoretical Explanations and the Role of Small Firm/Large Relationships', *Regional Studies*, vol. 33, pp. 367–77.

Storey, D.J. (1982), *Entrepreneurship and the New Firm*, Croom Helm, London.

Storey, D.J. (1994), *Understanding the Small Business Sector*, Routledge, London.

Storper, M. (1993), 'Regional "Worlds" of Production: Learning and Innovation in the Technology Districts of France, Italy and the USA', *Regional Studies*, vol. 27, pp. 433–55.

Storper, M. (1995), 'The Resurgence of Regional Economies, Ten Years Later: The Region as a Nexus of Untraded Interdependencies', *European Urban and Regional Studies*, vol. 2, pp. 191–221.

Wever, E. (1997), 'Clusters of High-Technology SMEs in a Small Homogeneous Country: The Netherlands', in D. Keeble and C. Lawson (eds), *Networks, Links and Large Firm Impacts on the Evolution of Regional Clusters of High-Technology SMEs in Europe*, ESRC Centre for Business Research, University of Cambridge, Cambridge.

Wever, E. (1998), 'The Dutch Case: The Randstad', in D. Keeble and C. Lawson (eds), *Regional Reports*, ESRC Centre for Business Research, University of Cambridge, Cambridge.

Williamson, O.E. (1975), *Markets and Hierarchies: Analysis and Antitrust Implications*, The Free Press, New York.

7 Collective Learning, System Competences and Epistemically Significant Moments

CLIVE LAWSON

7.1 Introduction

Despite the quite significant attention to innovation and learning found in recent economic writings, it is fair to say that this literature suffers from the conflation of different forms of learning. Learning, whether by individuals, firms, regions, industries or nations is largely treated in an undifferentiated fashion. But learning processes are not everywhere the same. Thus there is a need for some further conceptual clarification in this area. The first aim of this chapter is, in response, to separate and clarify some different processes of learning and argue that a conceptualisation of these processes, based upon the idea of system competences, is not only of great value but is as relevant to the region as to the firm. In so doing, we draw upon existing illustrative ideas of regional collective learning.

Questions concerning the nature of individual, firm and/or regional learning, i.e., ontological questions, immediately give rise to epistemological questions, concerning how we might come to know about such processes, especially once the social nature of learning is emphasised. The second half of this paper argues that certain phenomena, focused upon in the regional collective learning literature as conditions for learning, assume added significance for the researcher. More specifically, I argue that these phenomena prompt a cycling between tacit and discursive knowledge that provides, what may be termed, windows of epistemic access for the researcher.

7.2 Some Distinctions

Much of this recent interest in learning and innovation is motivated by the identification of new forms of competition, indicative of a move to a 'knowledge-based economy'. Price competition, within systems in which technology is given, is considered to be of less importance in this new (flexibly specialised/post-fordist) era. Instead, sustained competitive advantage results from the ability of firms to create, obtain and/or utilise knowledge more quickly than competitors. Thus, technological change is continually occurring at a rate that reduces (or obviates) the role of price competition. These developments gain a regional or local dimension as globalisation develops in that some, but not all, regional advantages disappear, thus making those advantages that are not undermined by globalisation (typically those involving knowledge and learning) relatively more important in explaining a firm's competitive position (Markusen, 1996). Furthermore, as globalisation increases the communicability of some types of knowledge, those types of knowledge that can more easily (or only) be transferred locally become relatively more important as a source of competitive advantage (Maskell and Malmberg, 1999).

In response to the above developments, various accounts have emerged that share a common emphasis upon the collective nature of learning. In particular, the idea that individuals must be members of some extended group or community of some sort (firm, division within a firm, regional milieu, nation, etc.) has been formulated in various guises. Some ambiguities and tensions in this literature are generally accepted and usually attributed to the infancy of the tradition (categories are simply expected to become more consistent as the amount of work grows; see for example the introduction to Cohen and Sproull, 1996, p. x). However, I want to argue that much ambiguity and confusion in the learning literature stems from a failure to distinguish two essentially quite different processes. Specifically, I want to make a distinction between two senses in which the 'collectiveness' of learning is emphasised. The first can be termed 'learning within an epistemic community', the second 'system learning'.

In the case of learning within an epistemic community, the focus is upon the way in which the individual learns *in virtue* of being a member of a particular community, that is, in virtue of the particular social position that he or she occupies within some collectivity.[1] Knowledge is settled upon through the interaction of the community's members. This interaction is enabled by use of existing ideas, languages and conceptions (acquisition of which is crucial for membership) in order to formulate (socially construct) knowledge. Indeed, any group or community must use an existing stock of concepts, languages or perspectives in order to develop new concepts, to innovate and to learn. Central to this form of learning is the manner in which situated individuals *transform* the existing ideas and conceptions with which they are confronted. This is not to say that knowledge is *merely* constructed. Reality does, so to speak, kick back, producing all manner of constraint upon the form of construction possible. But this external reality is conceptualised and learned about using the language and prior knowledge of that community. These phenomena, such as the existence of new technologies, competitive conditions, and other members of the group, provide the basis for the real *interaction* of the community, but this interaction is always manifest in some socially specific context using the conceptual materials to hand.[2] The crucial point is that it is the individual that learns, rather than some organisation or group, and that learning would not be possible were it not for the relationships in which that individual stands.

The second sense in which learning is understood as collective, system learning, is where enabling or successful routines become incorporated as part of what the system/group/organisation actually is. The emphasis here is not so much on what the individuals learn as on the processes by which the successes and failures that individuals experience become encoded into the routines and practices of the collectivity of which they are a part. Knowledgeable behaviour is made possible by some individual acting in accordance with or utilising the routines constitutive of the collective. Learning or problem solving by members is still important, but the emphasis is upon how this leads to a change in the constitutive routines of the community. Members of the community, by acting within the boundaries of what is acceptable or legitimate (in accordance with routines and procedures of the community) thus have access to the knowledge obtained by other members without either having access, necessarily, to the actual histories in which knowledge is obtained or consciously formulating their understanding of the processes involved. The knowledge is stored or encoded in the routines of the group, and it is the organisation or group that is portrayed as learning. Routines are the things encoded. Routines in this sense act much as, say, the rules of the Highway Code. Individuals, by serving an apprenticeship and becoming part of a particular community become familiar with rules that embody the learning of past members, enabling the new member to act capably, to safely travel from one destination to another. The structure of rules is constantly undergoing change as some elements are found to work better than others, as new technological needs/conditions emerge, and as the values and targets of the community changes. But access to the lessons of history is provided in a way that access to that history itself is not. The community is often viewed as having a 'collective memory', the store of routines, and so a form of intelligence. But the routines are not merely a 'stock'. Existing routines influence how individuals actually learn from direct experience, how such experience is interpreted, and how learning comes about from the experience and practice of others. It is not always clear what is learned by the individual agent. The emphasis is, instead, on the importance of tacit (procedural, uncodified) knowledge in explaining how agents act capably, rather than knowledgeably. However, the focus is not really upon the individual at all, but on the collectivity.

Both of the above categories of learning differ from (and must involve a rejection of) a simple 'transfer' conception which, isolating knowledge from practice, conceives of learning as the transmission of abstract knowledge from 'the head of the one that knows to the head of the one that does not'. Shared languages, knowledge, conceptions and often identities are crucial conditions for learning and knowledge acquisition to take place – underlining that learning is both active, as well as context dependent, *work*. The nature of social interaction within and between different social groupings is significant to collective learning, as are the differences in the forms of such groupings, especially in terms of their productive capabilities.

The Organisational Learning literature, as developed by amongst others Cyert and March (1963), and Nelson and Winter (1982), is perhaps the most prominent example of learning by a collective, in this case a firm. Of significance here is the fact that this literature serves to illustrate the problems that arise from a failure to make the distinctions noted above. Although general statements of the domain of organisational learning appear concerned solely with learning of the second type (system learning) (to illustrate this see especially the review paper by Levitt and March, 1996) the main ideas involve a reference to both types, mostly in an undifferentiated manner. This failure to distinguish quite different processes appears to underlie much of the confusion in this literature over the question of whether learning is really occurring – if learning requires knowledge and 'a knower'. Clearly agents act capably. And ways of doing things outlive the individuals that come and go within an organisation. But in what sense might this process be termed learning? Specifically, in what sense can we talk of the learning of collectives? In answering this question it is useful to consider the competence perspective.[3]

7.2.1 The Competence Perspective

The competence perspective has recently received a considerable amount of attention. Indeed, several recent volumes are completely given over to the task of clarifying the history of, and connections between, such terms as competences, capacities and capabilities (see for example Hamel and Heene, 1994; Montgomery, 1996; Foss and Knudsen, 1996). Although the precise meaning to these terms varies a little between authors, it is generally accepted that there exists sufficient family resemblances between the uses of these terms to talk of a coherent (and well established) 'competence perspective' (especially see Foss, 1996).

The common denominator in these accounts is the focus upon a potential that depends upon the way in which a something is structured. Bicycles can transport their riders because of the way they are structured, cups have the power to hold tea because of the way they are structured. Certain kinds of things, events and states of affairs such as the movement of cyclists or the non movement of tea, are explained in terms of other kinds of things, such as the arrangement of different tubes and chains, or the chemical composition of different metals or ceramics.[4] In the social realm, these structures consist in rules, relations and positions.[5] Two points should be made about these latter kinds of structure. First, none can be reduced to the actions they may govern. For example the refusal of a single motorist to stop at a red light does not negate our understanding of the rule 'when at traffic lights stop if the light is red'. Neither does it force us to consider such a rule as an 'average' or 'normal' description of what people do, even though most people may indeed stop at red lights. The rule is not, and cannot be evaluated as, a prediction of actual behaviour, it is something different in kind. Secondly, although structures are not actions, they nevertheless are an emergent feature of such action.[6] Thus these rules, relations and positions are not independent of human action. If the human race disappeared tomorrow, or simply stopped driving, the Highway Code would disappear in a way that features of the natural world would not.

Similarly, firms have potentials because of the way they are structured. For example, kinds of routines facilitate capable reactions to changing market conditions, to the emergence of new technologies and so on. The point is that not all of this capable behaviour, which involves knowledge, learning, or innovation, is best conceptualised (even metaphorically) as learning. But knowledge creation, transfer, and encoding can all usefully be understood as examples of the competences of productive systems such as firms that take place because of the social relations constitutive of different productive systems. The crucial distinction at the heart of the competence perspective is between the potential and actual characteristics of firms. Because of the way any organisation is structured, it has numerous competences, each of which can be realised/manifest in various ways (in products, market positions, and so on). Learning processes of individuals, knowledgeable behaviour that draws upon encoded routines, are simply different manifestations of such structures. As such, it makes sense to talk of 'system learning', as distinguished above, without attributing characteristics to the firm that pertain to individuals.

7.2.2 Collective Learning and Regional Competences

Now, if firms can indeed be meaningfully and fruitfully conceptualised in terms of their competences, it would seem that other social systems can also be so conceptualised to good effect. For example the region can be understood as systems of competences that include similar learning processes to those noted above, avoiding the idea that regions themselves somehow learn (as is encouraged by the Learning Region literature[7]). Both regions and firms are constituted by rules, relations and positions that are either reproduced or transformed through individual action. And both are emergent features of social activity. Of course, this assessment raises the question of what distinguishes regions from firms. Ultimately, the distinction between the two must become a substantive issue, resting upon the identification of the manner in which interaction, constitutive of the competence in question, is reproduced or transformed. In this, a crucial difference between the two (which other differences may often reduce to in practice) will be the relevance of contractual/legal rights and obligations. Firm competences are crucially constituted by interaction confined (or defined, along with membership, identity, etc. of those within the firm) by such contractual/legal considerations (the main insight of the contractarian tradition). However, these are not likely to have much direct bearing upon the kinds of interaction constitutive of regional competences. Thus, when focusing upon regional competences, there is a particular onus to account for any coherence (reproduction) of relationships observed. Furthermore, the relevant interaction will tend to take place between organisations and between different types of organisation. As such, the different means by which relationships emerge and are sustained between organisations becomes a central concern.

To return specifically to the issue of learning, regions differ from firms in involving interaction between individuals that are members of quite different epistemic communities. Individuals from suppliers, providers of financial or technical services, local support agencies other organisations are brought together, generating, if you like, communities across communities. Although there is at least the possibility of a greater benefit from the cross-fertilisation of ideas, issues of identity and the motivation for continued co-operation and interaction will doubtless be more conscious to those interacting, and more fluid than those within the firm.

Highlighting these distinctions between communities within the region raises the issue of whether there can really be any kind of system-learning at the regional level. In other words, what is the analogue, at the regional level, of the firm routines that become encoded with the lessons learned by members, and then act to facilitate the actions of other members? It is in regard to this question that the literature on collective learning, inspired by the work of the GREMI economists,[8] becomes significant. This literature focuses on the nature of a particular 'innovative milieu' within which firms operate and the kinds of learning that are made possible because of membership of such a milieu (see especially Camagni, 1991). For present purposes the main feature of these accounts is the concern with connections that facilitate collective learning. In particular, skilled labour mobility within the local labour market, various kinds of customer-supplier relations as well as relations to all kinds of service firms, and new start-ups through direct or indirect spin-out activity, feature as crucial channels whereby such learning takes place.

Now, just as knowledge is embedded in firm routines, so too knowledge is embedded in the structure of the labour market, in the structure of localised inter-firm relations and the institutional framework within which firms not only interact but come into being. Indeed these features become the main repositories for knowledge that enable system learning at the regional scale, underlying the nature of the competences of that region.

To recap, I have argued that different kinds of learning process can be identified at both the firm and regional level. The first relates to learning made possible by being part of some collectivity, the second to the way a collectivity encodes the lessons of its members in the structures that constitute it. I have also argued that such processes are usefully conceptualised in terms of system competences.

However, as suggested above, these ontological questions quickly give rise to epistemological questions, concerning how we might come to know about such processes. Emphasising the social character of learning and its dependence, at any point, upon a complex of rules, relations and positions underlines the difficulty of learning about such learning. In the remainder of this chapter I shall argue that phenomena focused upon in the regional collective learning literature (as indicators of the possibility of such learning), have a further significance with regard to coming to understand the kinds of learning process focused upon above. These arguments can be best organised by focusing upon the notion of epistemically significant moments.

7.3 Epistemically Significant Moments

Epistemically significant moments are moments in time that, because of their particular character, have a certain importance or status with respect to learning for either the agent involved, the social investigator of that act, or both. The main feature of these moments is some form of transition, crisis or rupture in the structural conditions of a practice, that prompt a 'reclaiming' of knowledge that the agent already has, but is not discursively aware of.[9] At root here is a distinction between different levels of consciousness. Following Giddens (1984), a discursive and tacit level of consciousness can be distinguished, giving rise to the possibility of 'reclaiming' knowledge in the sense that there is scope for the movement between levels of consciousness. The epistemic significance of such moments for the agents involved should be clear. However, I also want to argue that such moments are important to the researcher because they offer access to a (social) reality that is inherently interconnected and for the most part tacitly apprehended even by those agents most immediately involved. Any event that prompts agents to reflect on the web of relationships in which they stand or on the complex tacit knowledge that organises their activity provides a window of epistemic access onto the social reality under investigation. Individuals come to understand and to control their environments by being forced to formalise all manner of knowledge usually taken for granted. And the researcher, by focusing on these moments of rupture, is able to gain particular insight into whole areas that would otherwise be inaccessible. All that is needed is that some event or state of affairs provokes such cycling. Thus moments where tacit knowledge passes through a discursive or articulated stage are particularly significant to social research in a way that is clearly irrelevant to natural science.

Although the idea of an epistemically significant moment seems to be implicit in a diverse collection of writings, it is rarely explicitly discussed or developed.[10] An exception to this, which has the advantage of being set in an explicitly economic context, is provided in the work of J.R. Commons. In Common's work, much attention is given to the processes by which disputes are settled. There are various reasons for this focus by Commons including the fact that such settlement acts as the main mechanism whereby the interests of different groups are discovered, formalised and interpenetrate. But a crucial aspect of this procedure for Commons is the continual interaction between rules and routines laid down (or legitimised) by some central authority and those that emerge out of custom and practice. Of course the failure to make a polar separation between these is seen by (mostly Austrian) critics as a failure to understand the nature of theoretical research (in its Mengerian form at least), which should be preoccupied with the spontaneously emerging institutional forms (see Vanberg, 1989). However, on a more careful reading of Commons, it becomes clear that he is concerned with how these levels interact (see Lawson, 1994). In focusing upon such interaction, Commons emphasises that it is not simply the emerging or 'given' nature of these rules and routines that are important, but their tacit and discursive qualities. Although agents are aware of some rules more consciously than others, the importance for Commons is that whilst customs may come about tacitly in response to some repetitive situation where useful ways of acting may be copied, or subconsciously reproduced, the need to settle conflicts prompts a more conscious awareness of certain conditions, of their day to day action, that is, customs constantly move through a discursively conscious moment:

> these customary standards are always changing; they lack precision and therefore give rise to disputes over conflicts of interest. If such disputes arise, then the officers of an organised concern, such as a credit association, the manager of a corporation, a stock exchange, a board of trade, a commercial or labour arbitrator, or finally the courts of law up to the Supreme Court of the United States, reduce the custom to precision and add an organised legal or economic sanction (Commons, 1934 (1990), p. 72).

Commons is particularly interested in the means by which custom and practice define the positions from which agents transact and provide the grounds for dispute which become the object of attempted transformation. In so doing, the agents concerned reflect upon the nature of the rules and routines they follow and the relations in which they stand, making all sorts of customary (tacit) knowledge precise (discursive/articulated). Commons is explicit about how these moments serve to provide knowledge for the agent and for the social researcher and how these moments serve to incorporate knowledge (whilst at the discursive stage) into working procedures and norms of the particular authoritative organisation (in much the same way as rulings become precedents in the courts).

Now it should be clear that although Commons is only really concerned with situations of conflict, and especially those within the firm, as epistemically significant moments, the basic features of this account can be easily transposed to other processes that do not simply reflect some form of conflict. Importantly, these ideas can be transposed to processes that involve some form of conscious learning or change (innovation) and can be transposed outside the domain of the firm altogether.

An important example is provided by Nonaka and Takeuchi (1995).[11] Nonaka and Takeuchi focus upon just this cycling between articulated and tacit knowledge, but in the context of new product development. They also see this cycling as consisting in four distinct stages. The first involves sharing tacit knowledge among organisational members. This refers not only to shared values and norms and to a shared capacity to understood the codes in which knowledge is articulated, but also to shared technical knowledge. The second stage is when individuals with diverse and complementary knowledge come together and collectively seek to articulate their ideas about a new product or technology. As individuals try to articulate intuitions or rough ideas about a new product or a technology, they are forced to clarify their ideas and to develop new and more adequate concepts or models about the technology they are trying to develop. In its altered and more explicit form, knowledge becomes easier to combine with other known technologies and methods that may be incorporated into a new product. This process of combining explicit knowledge is the third stage in the innovation cycle. It culminates in the building of a prototype, which can be tested and modified in relation to established evaluation criteria. The fourth stage involves the movement from articulated back to tacit knowledge. The idea here is that as the organisation produces the new product in large quantities and becomes proficient in the new technology, much of the underlying knowledge needed for its production becomes tacit. Routines emerge and the members act in a co-ordinated and capable way without needing to explain in words or diagrams exactly how they are able to do so. This establishes an altered base of tacit knowledge upon which new knowledge may be created.

According to Nonaka and Takeuchi the second stage is the critical one because it is at this point that actors seek to make explicit ideas and notions about new products and techniques that previously only amounted to hunches or rough intuitions. Forcing the tacit into the discursive allows for a creative *recombination and reconstitution of existing ideas*. Thus, as with Commons, the crucial moment is seen to be this point of transformation of tacit into discursive knowledge, but the emphasis in Nonaka and Takeuchi's account is upon the resulting ease of recombination of new ideas, and the effect of this on innovation.

The clear separation of this process into successive stages in this account also has the advantage that it facilitates the extension of the cycling idea to the regional level. Stage one is concerned with the existence of shared tacit knowledge, rules, routines, and languages, amongst the members of the regional community. Indeed, much of the GREMI literature on collective learning within regions (noted above) is very much concerned with the establishment of conditions for dialogue, between firms and between firms and other agencies. All the features emphasised in these accounts (intra-regional labour mobility, corporate and academic spin-offs, supplier-based relations) act to create this common tacit knowledge.

The next stage is concerned with the articulation of this tacit knowledge. This will tend to involve the articulation of the idea underlying a particular project. But it will also involve the articulation of the competences and capabilities of each firm prior to collaborative arrangements or networking activity. It may also involve articulating the basic idea underlying the start-up or spin-off of a new firm prior to approaching a funding agency. In a similar manner, this articulation makes possible the construction not only of a prototype, but the emergence of a new firm or set of collaborative arrangements, consortia, or partnerships. This stage, as brought out in Nonaka and Takeuchi's account, is crucially a stage at which 'combination' is possible because certain ideas and knowledge, are 'made precise'. This combination leads to a new thing such as a product, firm, or collaborative arrangement. As the focus becomes the production of the good, the working of that firm or the operation of the collaboration, this acts to return the idea that underlay its inception to the tacit consciousness of the individual agents involved. These ideas become latent in the routines and practices of firms and other agencies constitutive of the region.

The most significant moment here is, as for Commons and for Nonaka and Takeuchi, the movement from stage one to stage two. There are several routes by which this may occur. This movement may not simply come about as part of the innovation process, but as the lead up to a start-up or spin-off of a new firm. Also labour mobility will prompt this movement. Thus although not a part of one innovation process, individuals thinking about moving between firms and thinking about 'fitting in' to new firms, will be seeking to make precise all manner of routines, rules, and ideas.

This account brings together many of the elements introduced above. Commonly-held tacit knowledge is often pervasive in successful high-technology regions. In some cases this knowledge is unique to particular product areas, having emerged from a rich history of local interaction between users and producers of the technology. Often this emergence involves a fortuitous element, reflecting the coincidental presence of strong expertise in complementary producer and user industries. In other cases the critical factor may be the way the multidisciplinary culture of a local university, combined with a history of spin-offs, serve to diffuse it widely amongst local producers.[12] Such diffusion may take place because of important local cultural institutions, as the industrial district literature argues, or because of various forms of inter-firm collaboration or labour mobility as identified by GREMI regional economists and others. However, such mechanisms for the transference of the conditions of learning (such as spin-offs, or labour mobility) have the added significance of laying the conditions for the cycling between tacit knowledge and articulated knowledge that Nonaka and Takeuchi have argued is crucial to new product development.

At a formal level perhaps the central difference between the way this cycling process unfolds in regional settings as compared to organisational ones concerns the mechanism whereby rules and procedures are sanctioned or legitimised. In the case of decentralised regional communities of producers there is, of course, no direct parallel to the sanctioning power exercised by management over rules and procedures. Rather, what seems to bear most directly on the legitimacy of particular practices and arrangements is the way community opinion can enforce certain social norms or codes of behaviour, especially those that bear on acceptable forms of co-operation and knowledge exchange. The importance of community norms in enforcing a particular balance between co-operation and competition that contributes to a region's innovative capacity is, of course, one of the key insights of the industrial district literature.[13] Incidentally, this difference, which depends ultimately upon the manner in which different sets of relationships are reproduced or transformed, seems to be at the heart of various attempts to distinguish organisational from regional productive systems, especially those in terms of hierarchies and networks (see Thompson et al., 1993).

Two last comments ought to be made in the light of this discussion. First, this process of cycling brings together and clarifies the point of contact between the two types of learning distinguished in the earlier part of this paper. This kind of cycling is one process that incorporates both types of learning as different moments. Differences lie in the different focus upon particular start and end points of the cycle. For the learning in epistemic communities the start and end is the discursive knowledge that resides in the individual. With the latter, system learning, the start and end of the cycle is the tacit knowledge that resides in the routines of the organisation. The second point is that phenomena such as labour mobility, spin-offs and the undertaking of inter-firm relations (the usual focus of GREMI-inspired regional collective learning accounts) do not simply serve as indicators of the amount of learning that takes place within some regional productive system but as epistemically significant moments. These moments, in making tacit knowledge precise and explicit facilitate both the recombination of existing ideas into new ideas and provide the regional economist with access into the network of relations that enable learning and constitute the competences of that regional system. As such, these phenomena acquire the status of special objects of study that are likely to require detailed and sustained study in their own right.

7.4 Concluding Remarks

The main aim of this chapter has been to distinguish and clarify different processes of learning that are often conflated in the literature. It has also been argued that the processes of learning distinguished are most usefully conceptualised in terms of competences, thus avoiding some of the ambiguities and category mistakes that can be found in the existing firm and regional literature. This discussion serves to underline the importance of understanding those mechanisms by which shared knowledge, languages and cultures are diffused at both the firm and regional level. However, this account also highlights the importance of moments of rupture in routines and practices. Such moments not only have a crucial role to play in the development of new ideas and products, but also provide windows of epistemic access for social researchers. As such, phenomena such as firm disputes or (threat of) take-over at the firm level, or labour mobility, spin-outs and firms' decisions to undertake joint ventures and collaborations at the regional level, take on a new significance for the regional researcher.

Taking these points together, it would seem that the relative successes of both firms and regions may well depend upon sustaining a fruitful trade-off between those factors that provide continuity in reproducing or transmitting shared knowledge (the learning of 'routines' which enable capable behaviour) and those factors that disrupt, thus forcing tacitly-held knowledge to go through moments in which such knowledge is articulated and recombined.

Notes

1 The term 'epistemic community' has recently acquired quite specific (but not necessarily consistent) meanings in a number of disciplines (especially in international relations (Adler and Haas, 1992; Zald, 1995) and feminism (Nelson, 1993; Assiter, 1996)). However, I am using the term here more generally, to draw attention to the fact that knowledge is always acquired by individuals within, and conditioned by, sets of social relations.

2 In short, an ontological realism is combined with an epistemological relativism. The most fully elaborated philosophical account, and defence, of this position is that provided by Bhaskar (1989), see also Lawson (1997). For an extended substantive account of the processes by which membership occurs and conditions learning in relationship to such issues as apprenticeships, see Lave and Wenger (1991).

3 It is not possible to develop the arguments in favour of a general competence perspective in this context, but see for example the contributions in Foss and Knudsen (1996) and Lawson (1999).

4 Such examples are of course very static and are thus more directly examples of capacities than of competences, where the latter carry the connotation of being acquired in some way (and are thus more relevant to the social world).

5 For a more extended discussion of the kind of social ontology that a competence perspective presupposes, see Lawson (1999).

6 Some level of organisation can be said to be emergent if there is a sense in which it has arisen out of some lower level but is not reducible to it or predictable from it. For present purposes, the two features of this conception of most importance are that the higher level is not independent of the lower level out of which it has arisen, and indeed is conditioned and 'rooted in' this lower level, and that the highest level cannot be predicted from the lower if it really is emergent in any real sense. Neglect of the former leads to an omission of the human-dependent nature of social structures, neglect of the latter leads to no real notion of social structure at all (as witnessed by even the more sophisticated methodological individualist accounts). For a more detailed discussion of this notion in the economic sphere see Lawson (1997).

7 See for example Asheim (1996), Morgan (1997) and Simmie (1997).

8 It should be pointed out that the term collective learning appears in a wide range of contributions and that the meaning attributed to the term is often quite vague, usually being understood as a quite general counterpart to a notion of *individual* learning. The term has been used in relation to the idea of core competences (Prahalad and Hamel,

1990), and firm specific competences (Pavitt, 1991, p. 42), and as part of a general discussion of organisational learning (Dodgson, 1993). However, the term acquires a distinctly regional orientation, and some precision, in the work of the GREMI economists (see especially Camagni, 1991).
9 To be precise, two essentially different processes are set in motion from such crises or ruptures. Along with the 'reclaiming of the tacit' which is the main focus in the remainder of this paper, is what has been referred to as 'the methodological primacy of the pathological' (Collier, 1977, p. 132). By seeing how something goes wrong we can find out more about the conditions of its working properly than we ever would by observing it working properly. An economy or firm in crisis or undergoing some momentous change is more 'transparent' than a normally functioning one (Collier, 1994, p. 164).
10 One exception to this is the discussion provided by Bhaskar (1989, p. 48). However, the emphasis in Bhaskar's account is upon the 'primacy of the pathological' rather than the reclaiming of the tacit.
11 For a more detailed account of the relevance of Nonaka and Takeuchi's ideas in this context, see Lawson and Lorenz (1999).
12 See Lawson and Lorenz (1999).
13 See, for example, Brusco (1986) and Lorenz (1992).

References

Adler, E. and Haas, P.M. (1992), 'Epistemic Communities, World-Order, and the Creation of a Reflective Research-Program – Conclusion', *International Organization*, vol. 46, pp. 367–90.
Asheim, B. (1996), 'Industrial Districts as "Learning Regions": A Condition for Prosperity?' *European Planning Studies*, vol. 4, pp. 379–400.
Assiter, A. (1996), *Enlightened Women: Modernist Feminism in a Postmodern Age*, Routledge, London.
Bhaskar, R. (1989), *The Possibility of Naturalism*, Harvester, Brighton.
Brusco, S. (1986), 'Small Firms and Industrial Districts: The Experience of Italy', *Economia Internazionale*, vol. 39, pp. 98–103.
Camagni, R. (ed.) (1991), *Innovation Networks: Spatial Perspectives*, Belhaven Press, London.
Cohen, M.D. and Sproull, L.S. (eds) (1996), *Organizational Learning*, Sage, London.
Collier, A. (1977), *R.D. Laing*, Harvester Press, Hassocks.
Collier, A. (1994), *Critical Realism: An Introduction to Roy Bhaskar's Philosophy*, Verso, London.
Commons, J.R. (1934) (1990), *Institutional Economics: Its Place in Political Economy*, Transaction, New Jersey.
Cyert, R.M. and March, J.G. (1963), *A Behavioural Theory of the Firm*, Prentice Hall, Englewood Cliffs.
Dodgson, M. (1993), 'Organisational Learning: A Review of Some Literatures', *Organisation Studies*, vol. 14, pp. 375–94.

Foss, N.J. (1996), 'Higher-order Industrial Capabilities and Competitive Advantage', *Journal of Industry Studies*, vol. 3, pp. 1-20.
Foss, N.J. and Knudsen, C. (eds) (1996), *Towards a Competence Theory of the Firm*, Routledge, London.
Giddens, A. (1984), *The Constitution of Society: Outline of the Theory of Structuration*, Polity Press, Cambridge.
Hamel, G. and Heene, A. (eds) (1994), *Competence-Based Competition*, John Wiley & Sons, New York.
Hedberg, B. (1980), 'How Organisations Learn and Unlearn', in *Handbook of Organisational Design*.
Howells, J. (1996), 'Tacit Knowledge, Innovation and Technology Transfer', *Technology Analysis and Strategic Management*, vol. 8, pp. 91-106.
Lave, J. and Wenger, E. (1991), *Situated Learning: Legitimate Peripheral Participation*, Cambridge University Press, Cambridge.
Lawson, C. (1994), 'The Transformational Model of Social Activity and Economic Analysis: A Reinterpretation of the Work of J.R. Commons', *Review of Political Economy*, vol. 6, pp. 186-204.
Lawson, C. (1999), 'Towards a Competence Theory of the Region', *Cambridge Journal of Economics*, vol. 23, pp. 151-166.
Lawson, C. and Lorenz, E. (1999), 'Collective Learning, Tacit Knowledge and Regional Innovative Capacity', *Regional Studies*, vol. 33, pp. 305-17.
Lawson, T. (1997), *Economics and Reality*, Routledge, London.
Levitt, B. and March, J.G. (1996), 'Organizational Learning', in Michael D. Cohen and Lee S. Sproull (eds), *Organizational Learning*, Sage, London, pp. 516-41.
Lorenz, E.H. (1992), 'Trust, Community and Co-operation: Towards a Theory of Industrial Districts', in M. Storper and A.J. Scott (eds), *Pathways to Industrialisation and Regional Development*, Routledge, London, pp. 195-204.
Markusen, A. (1996), 'Sticky Places in Slippery Space: A Typology of Industrial Districts', *Economic Geography*, vol. 72, pp. 293-313.
Maskell, P. and Malmberg, A. (1999), 'Localised Learning and Industrial Competitiveness', *Cambridge Journal of Economics*, vol. 23, pp. 167-85.
Montgomery, C. (1996), *Resource-Based and Evolutionary Theories of the Firm: Towards a Synthesis*, Kluwer, Boston.
Morgan, K. (1997), 'The Learning Region: Institutions, Innovation and Regional Renewal', *Regional Studies*, vol. 31, pp. 491-503.
Nelson, L. H. (1993), 'Epistemological Communities', in L. Alcoff and E. Potter (eds), *Feminist Epistemologies*, Routledge, London, pp. 121-60.
Nelson, R., and Winter, S. (1982), *An Evolutionary Theory of Economic Change*, Harvard University Press, Cambridge Mass.
Nonaka, I. and Takeuchi, H. (1995), *The Knowledge Creating Company*, Oxford University Press, New York.
Pavitt, K. (1991), 'Key Characteristics of the Large Innovating Firm', *British Journal of Management*, vol. 2, pp. 41-50.
Polanyi, M. (1966), *The Tacit Dimension*, Routledge, London.
Prahalad, C.K. and Hamel, G. (1990), 'The Core Competence of the Corporation', *Harvard Business Review*, vol. 68, pp. 79-91.

Senker, J. (1995), 'Tacit Knowledge and Models of Innovation', *Industrial and Corporate Change*, vol. 4, pp. 425–47.

Simmie, J. (ed.) (1997), *Innovation, Networks and Learning Regions?*, Jessica Kingsley, London.

Teece, D., Pisano, G. and Shuen, A. (1997), 'Dynamic Capabilities and Strategic Management', in N. Foss (ed.), *Resources, Firms and Strategies*, Oxford University Press, Oxford.

Thompson, G., Frances, J., Levacic, R. and Mitchell, J. (eds) (1993), *Markets, Hierachies and Networks*, Sage, London.

Vanberg, V. (1989), 'Carl Menger's Evolutionary and John R. Commons' Collective Action Approach to Institutions: A Comparison', *Review of Political Economy*, vol. 1, pp. 336–62.

Zald, M.N. (1995), 'Progress and Cumulation in the Human-Sciences After the Fall', *Sociological Forum*, vol. 10, pp. 455–79.

8 Collective Learning Processes in European High-Technology Milieux

DAVID KEEBLE

8.1 Conceptualising Regional Collective Learning

The preceding chapter has examined particular conceptual issues associated with disentangling theoretically what is involved in the concept of 'collective learning', whether applied at the firm or regional level. One of its key conclusions is that this must involve attempting to understand 'those mechanisms by which shared knowledge, languages and cultures are diffused at ... the regional level'. This chapter takes up this issue, and attempts to review and synthesise the detailed empirical evidence generated by the European network's research on the nature and extent of regional collective learning mechanisms or processes operating in the European high-technology SME clusters studied. A key aim is to assess what particular and identifiable processes are important for collective learning, and how far these clusters can be viewed as 'learning regions' (see chapter 1) characterised by active and effective processes of regional collective learning which lead to continuing SME innovation and the development of new technology-based products and services.

With these objectives, the theoretical context for this chapter and the research conducted by members of the network is provided by two separate but related bodies of work, one concerned with learning processes in regional development, the other with learning processes within the firm. Theoretical understanding of the development of technologically-dynamic regions in Europe and North America has been greatly enhanced over the past decade by workers such as Scott (1988) and Storper in the United States and the GREMI group of European researchers in Europe. Storper's work (1997, ch. 6) on technology-based 'regional worlds of production', or 'knowledge communities' as Henry and Pinch (2000) prefer to call them, has drawn attention to the key role in their growth of 'untraded inter-dependencies' between local firms and other organisations (Storper, 1995). Unlike earlier workers' preoccupation with measurable input-output flows of goods and products, Storper stresses the importance for firm innovation and growth in technology districts of less easily measurable forms of localised interaction and inter-dependencies involving informal inter-firm networking and collaboration (Yeung, 1994) and processes of regional 'collective learning' (Camagni, 1991a; Lorenz, 1992; Lazaric and Lorenz, 1997; Lawson, 1997a, 1999). These processes, which result in the exchange and development of organisational and technological expertise and high rates of technological and product innovation, are seen as being particularly characteristic of regionally-clustered high-technology SMEs, such as those in Silicon Valley or Orange County, California.

Storper's theoretical work has been paralleled in Europe by that of the GREMI (Groupe de Recherche Européen sur les Milieux Innovateurs) school of regional economic research (Aydalot, 1986; Aydalot and Keeble, 1988; Camagni, 1991b; Ratti et al., 1997). It is this group which has explicitly developed the concept of 'collective learning' to connote a broad notion of the capacity of a particular regional 'innovative milieu' to generate or facilitate innovative behaviour by the firms which are members of that milieu. Indeed, for Camagni (1991a, p. 130), collective learning is central to the development and definition of a successful milieu; 'the local 'milieu' may be defined as a set of territorial relationships encompassing in a coherent way a production system, different economic and social actors, a specific culture and a representation system, and generating a *dynamic collective learning process*'. This collective learning process involves regional mechanisms which reduce the uncertainty faced by firms in a rapidly-changing technological environment (Lawson, 1997a), and is seen by Camagni (1991a, p. 127) as operating 'mainly through skilled labour mobility within the local labour market, customer-supplier technical and organisational interchange, imitation processes and reverse engineering, exhibition of successful 'climatisation' and application to local needs of general purpose technologies, informal 'cafeteria' effects, complementary information and specialised services provision'. Camagni's conceptual-isation of regional collective learning thus focuses on links and networking between firms and via the regional labour market, accords it a central role in the development of a successful innovative milieu, and pinpoints a number of key mechanisms by which it may take place. It is important to note that these include both 'conscious' and 'unconscious' mechanisms (Capello, 1998), an example of the former being deliberate research collaboration between local SMEs or between an SME and a local university, examples of the latter being local skilled worker mobility or entrepreneur spin-off from existing local firms to create new technology-intensive firms. Both conscious and unconscious mechanisms involving interaction and diffusion of knowledge between firms and organisations may thus generate a regional collective learning capability which sustains continuing innovation by the cluster's firms.

The second body of literature which provides theoretical insights relevant to analysing empirically processes of regional collective learning is that on learning processes within the firm (for example March, 1991), as reviewed and developed by Lorenz (1996; Lazaric and Lorenz, 1997; see also Lawson, 1997a, 1999; Lawson and Lorenz, 1999; Maskell and Malmberg, 1999). This literature is concerned with the ways in which firms seek to overcome internal coordination problems by constructing shared knowledge in the form of commonly understood rules and accepted procedures. By extension, regional collective learning is defined by Lorenz (1996) as involving 'the creation and further development of a base of common or shared knowledge among individuals making up a productive system which allows them to co-ordinate their actions in the resolution of the technological and organisational problems they confront'.

In this context, Lorenz identifies three areas in which firms need to develop shared knowledge. First, in terms of preconditions for learning, there is the need to establish a common language for talking about technological and organisational problems. This is closely related to the need for common standards of honesty and information sharing as the basis for the adaptation of industrial partners to unanticipated contingencies not explicitly provided for in formal contracts. As Lorenz (1996) points out, 'a clear understanding and mutual consensus over the rules provides a basis for the progressive build-up of trust, which is arguably indispensable for innovative collaboration, given the uncertainties which surround its terms and outcomes'.

Secondly, 'there is a need for a shared knowledge of a more strictly technological or engineering sort, which allows different firms to effectively collaborate in a technological project' (Lorenz, 1996). This knowledge is not simply (or most importantly) concerned with core research, but with the more down-stream phase of innovation, involving detailed product design, testing, re-design and production. This 'in-house' knowledge is often difficult to transfer because it is not easily codified as 'its transfer depends ultimately on the mobility of individuals or teams' with practical experience in the technology concerned. The third kind of shared knowledge is organisational, examples suggested by Lorenz being how to manage hierarchical relations, how to divide responsibilities among different occupations or services, or what procedures are needed to assure the consistency of collective decision-making.

Lorenz's approach, though from a different theoretical starting point, thus bears a number of similarities to that of Camagni and the GREMI. There is the same stress on the need for firms to reduce uncertainties by sharing and collaborating, the same implicit emphasis on local inter-firm relations or networking, and the same recognition of the probable importance of such mechanisms or processes of regional collective learning as the movement of key research staff or entrepreneurs between firms; as Lorenz (1996) stresses 'mobile workers [are] the carriers of knowledge on the local labour market'. Equally, Camagni like Lorenz recognises the importance of preconditions for learning, in the form of common 'tacit codes of conduct ... and the formation of common 'representations' and widely shared 'beliefs' on products and technologies'. This preconditions aspect of regional collective learning Camagni sees as likely to be encouraged by 'synergy effects stemming from a common cultural, psychological and often political background, sometimes enhanced by the effectiveness of some local 'collective agent' (Camagni, 1991a, pp. 133–4). Camagni's longer theoretical discussion is however more explicit about the key role of geographical proximity in the development of collective learning, stressing as it does the role of locally-rooted (at least to some extent) human capital resources whose 'presence accounts for much of the local collective learning process', the 'presence of an intricate network of mainly informal contacts among local actors, building what Marshall called an 'industrial atmosphere', made up of personal face-to-face encounters, casual information flows, customer-supplier co-operation and the like' (Camagni, 1991a, p. 133), and the local synergy effects associated with a common cultural background noted above.

The above discussion thus pinpoints several attributes of and key processes likely to be operative within a 'high-learning' regional milieu characterised by an active collective learning capability. First, such a regional milieu can only develop if appropriate socio-cultural preconditions for learning involving different firms and organisations are present. Second, regional collective learning is strongly associated with the movement of key individuals and skilled workers, carrying technological and managerial knowhow and 'embodied expertise' as an unconscious learning mechanism, between local firms and other organisations. This can clearly occur in at least two ways, namely local labour market recruitment of skilled workers, and new firm spin-off where an entrepreneur or entrepreneurs[1] take ideas, expertise or potential products which they have developed in a 'parent' firm or organisation and establish a new local business to further develop and exploit them. In both cases, the individuals concerned often retain continuing links and personal contacts with their previous employing organisation through which ideas and knowledge can continue to diffuse, interact and develop.[2] Third, a collective learning capacity is likely to be strongly enhanced by high levels of conscious networking and collaboration between local firms and organisations, both small and large. Finally, the existence of active regional collective learning processes and of a 'high-learning' regional milieu is likely to be evident in high rates of firm product and service innovation, as a key output measure of the effectiveness of local learning processes. Each of these potentially measurable indicators of the existence of an effective collective learning capability in the European high-technology clusters studied by the network will be considered in turn.

8.2 Preconditions for Collective Learning

One of the most important, but also most elusive and difficult to measure, aspects of regional collective learning capacity stressed by both Camagni and Lorenz is the need for pre-conditions for learning, in terms of common regional culturally-based rules of behaviour, language of engagement and collaboration, and accepted but tacit codes of conduct between firms, which enable the development of trust, itself essential for innovative collaboration. As Camagni notes, the development of these 'cultural' pre-conditions may also be enhanced by the effectiveness of some local 'collective agent'.

Notwithstanding the heterogeneity of the study regions (see chapter 2), the network's research does identify certain types of regional 'collective agent' which in particular regions have clearly helped to create preconditions for the evolution of a regional culture of SME interaction and collective learning. The most obvious are major universities characterised by either liberal or technological cultures, but major public sector research institutes, and large private technology and R&D consultancies, also appear to play an important role in certain cases. The argument here is that major universities in particular may help shape a local culture amongst research-based businesses in which research interaction, dissemination, debate and collaborative endeavour are positively valued and widely practised. This culture thus results in frequent innovative and cross-fertilising research within and between local firms.

It is also possible that particular enlightened locally-based large firms, such as Siemens in Munich, Nokia in Helsinki and Oxford Instruments in Oxford, have through positive subcontracting relationships with local SMEs been influential in creating a regional culture of trust and collaboration which encourages innovation. It is noteworthy that with the possible exception of Göteborg, none of the study regions have been characterised by regional industrial cultures dominated by dependency relationships controlled by nineteenth-century monolithic industries and firms (coal, steel, shipbuilding, textiles), and that, as de Bernardy (1999, p. 349) notes in relation to Grenoble, their labour markets nearly all possessed some initial bias in socio-cultural terms towards highly-qualified 'knowledge workers' or professionals.

In general, nearly all the study regions have thus historically been preconditions-rich rather than preconditions-poor with respect to the development of regional collective learning processes. Possible exceptions to this generalisation are Barcelona (notwithstanding well-developed universities) and Sophia-Antipolis (created in a socio-cultural vacuum), while Randstad SMEs appear to operate within a national rather than regional socio-cultural environment. With respect to the regional typology set out in chapter 2, effective regional collective agents encouraging the development of preconditions for learning have perhaps most obviously been present in the three University-based regions studied (Cambridge, Oxford and Pisa). But they undoubtedly also exist in both Industrial and Metropolitan regions, as illustrated by Table 8.1.

A final and very important point arising from the network's case study research is that establishing appropriate preconditions for collective learning, even where a potentially effective regional collective agent exists, is heavily time-dependent. De Bernardy's (1999, p. 351) account of the development of collective learning in Grenoble, for example, stresses the fifty-year time span which has been needed, notwithstanding the activities of its local collective agents:

> this capacity for collective learning, much of it unconscious rather than formalized, represents a common culture of the city of Grenoble, a culture that over the generations has demonstrated its capacity to produce change by an original mix of individual social and professional investments and of

Table 8.1 Pre-conditions for learning: examples of regional collective agents

Cambridge

1) Cambridge University – generally liberal and positive attitude towards research collaboration, sharing and development, which may have spilled over into and helped shape a wider culture of innovative interaction within the local research-based firm community via university spin-offs, researcher recruitment, research collaboration ... 'the Cambridge approach stands in sharpest contrast to [that] of most other British Universities [involving] a central perception of the strategic value of industrial links and a commitment to its realisation ... through a reliance on research excellence and on liberal ground rules governing its exploitation rather than by means of formal regulation and institutional devices' (Segal Quince Wicksteed, 1985).

2) Large locally-founded technology and R&D consultancies (Cambridge Consultants, PA Technology, Scientific Generics, The Technology Partnership), which appear to have generated and fostered numerous local research-intensive spin-offs, in a pro-active and positive fashion, encouraging a local culture of trust and technological collaboration. For example, 'Cambridge Consultants ... launched in 1960 by a group of newly graduated scientists and engineers ... has exercised a distinctive influence on the Cambridge high technology business scene, directly through the number of companies it has helped spin out from itself and indirectly ... as indicative of the creativity and individual enterprise of the University's engineers and scientists' (Segal Quince Wicksteed, 1985): 'Absolute Sensors ... is one of a string of spin offs in recent years from Scientific Generics, the technology consultancy ... Generics provides basic housekeeping functions such as IT support, accountancy and an extremely pleasant environment for a start-up venture' (Chapman, 1999).

Grenoble

1) Public sector research institutes, such as CEA-G (nuclear research), INPG (engineering research and training), and LETI and CNET (microelectronics): the first two also established new firm incubators (ASTEC and Hitella). 'Employing more than 2,500 technicians, engineers and scientific researchers linked to the university, CEA-G was from its outset committed to local firm collaboration ... In the same way, INPG developed numerous partnerships with local firms and was the source of local

computer firm spin-offs from a very early stage ... These institutions have played a very important role in new firm formation' (de Bernardy, 1999, p. 346). 'The Grenoble Network Initiative (GNI) is a network linking ... local firms with participation from local public sector research institutes and agencies. In addition to establishing a communication network, GNI provides workshops to improve training, internet and electronics applications, and co-operation between the health and electronics sectors. In only a few years it has had a significant impact on local innovative activity and has strengthened the innovative milieu' (de Bernardy, 1999, p. 350).

2) Grenoble's universities: AUG (University Alliance of Grenoble) established 1947, many links with local firms and public research institutes: 'AUG still plays a catalyst's role in generating synergetic and symbiotic effects and creating a culture for co-operation between complementary business activities' (de Bernardy, 1999, p. 346), while 'Grenoble's universities have played a growing role' in strengthening the innovative milieu, via research clubs, formal research co-operation, and student placements (de Bernardy, 1999, p. 350).

Helsinki

1) Public research institutes and technology transfer agencies: VTT (Technical Research Centre of Finland), located next to Helsinki University of Technology in the Otaniemi Science Park (latter 'the largest centre of technological knowledge and business in Northern Europe'); TEKES, Technology Development Centre of Finland, and SITRA, Finnish National Fund for R&D (largest venture capital fund in Finland): 20 government research institutes in the Helsinki Region.

2) Universities (8 in Helsinki Region), especially University of Helsinki (32,000 students) and Helsinki University of Technology (12,500 students): liberal innovation exploitation rules for new firm spin-offs (Kauranen, 1997, pp. 8–10).

organized public initiatives, particularly in fostering research capabilities and support structures.

Only in the later 1990s, after 25 years of evolution, have new technology clusters such as Sophia-Antipolis begun to develop appropriate preconditions through the proactive efforts of local public research institutes and the University of Nice/Sophia-Antipolis. The development of collective learning capabilities is thus heavily time-dependent, even where favourable preconditions may exist.

8.3 New Firm Spin-Offs as a Collective Learning Process

The theoretical discussion of section 8.1 stressed the importance as a regional collective learning mechanism of the movement of key individuals and skilled workers, carrying technological and managerial knowhow and 'embodied expertise', between local firms and other organisations. One very important type of movement in this respect is via new firm spin-off, where entrepreneurs with research, engineering, scientific or managerial knowhow take ideas, expertise or potential products which they have developed in a 'parent' firm or organisation and establish a new local business to further develop and exploit them. In leaving their existing firm, university or other local organisation to set up a new firm to exploit a new technology, innovation or market opportunity, spin-off entrepreneurs diffuse within the region high-level expertise and competences, thereby augmenting and developing the local pool of knowledge. High-technology spin-offs are also frequently created by two or more founders,[3] bringing together different but complementary skills and knowledge, again often resulting in new knowledge development and collective learning. Such entrepreneurs also often retain close links with their 'parent' organisation (Lindholm Dahlstrand, 1999), creating opportunities for networking, collaboration and the development of yet further 'untraded interdependencies' (Storper, 1995), while their new firms in turn often spin-off further second and third-generation enterprises, in a cumulative and dynamic process of knowledge diffusion and development. A high rate of new firm spin-off and resultant transfer of 'embodied expertise' within regional clusters of technology-intensive SMEs thus represents a very important and dynamic process of regional collective learning.

The network's research provides powerful evidence of the central importance of new technology-based spin-offs, viewed here as key indicators of an evolving regional collective learning capability, in all the successful high-technology regional clusters for which data are available (Table 8.2).

Table 8.2 reveals that all the regions listed are now characterised by a dynamic but highly localised spin-off process except Barcelona, that spin-offs originate from both small and large firms, universities and public

Table 8.2 Technology-intensive spin-offs as a regional collective learning process

- **Cambridge** – 88% of high-technology SMEs are spin-offs or new start-ups, and 81% of these were founded by entrepreneurs formerly working for a local firm (56%) or university (19%): 48% of SMEs report further subsequent spin-offs from their own firm, all of which have remained in the Cambridge region, 75% with continuing links with 'parent' SME (Keeble et al., 1999).
- **Oxford** – 80% of technology-based SMEs are spin-offs or new start-ups, with 84% of founders coming from the Oxford region (42% from other firms, 29% from university): 44% report further spin-offs from their own firm, 86% of which are located in the Oxford region (Lawton Smith, 1997a, 1997b).
- **Göteborg** – 73% of new technology-intensive SMEs are spin-offs (idea originated in 'parent' firm or university), while 83% of founders of these firms are from the Göteborg region: large firms are the most frequent source of spin-offs, with Chalmers University of Technology second: 73% of firms retain links (especially customer links) with 'parent' (40% after 10 years) (Lindholm Dahlstrand, 1999).
- **Pisa, Piacenza, NE Milan** – 57% of 63 high-technology SMEs surveyed (all manufacturing firms) report being spin-offs, with 89% of spin-offs coming either from local firms (72%) or universities (17%): 'spin-off mechanisms [are] a distinctive element' of these technology-based clusters, while 'the attitude of entrepreneurs towards spin-off firms is extremely positive' (Campoccia, 1997, pp. 28–9).
- **Sophia Antipolis** – new spin-off phenomenon in 1990s, 50 direct spin-offs (out of c.400 firms?) by 1996, more now: most from public research institutes (INRIA, INSERM, Ecole des Mines) and University of Nice/Sophia-Antipolis, but also growing number of new start-ups by entrepreneurs leaving big firms (Thomson, DEC, Texas Instruments): result is 'multiplication of small and micro firms, with strong links with research and close interrelationships, created by 'Sophipolitan' entrepreneurs' (Longhi, 1999, p. 341).
- **Grenoble** – 'Since 1980, the dynamic of new firm creation has become very important for the region', with new micro-firms (under 10 employees) accounting for 66% of gross job creation, and all of net job creation, 1975–92, in private sector enterprises (de Bernardy, 1999, pp. 346–7).
- **Barcelona** – only weak local spin-off process thus far: only 26 university-origin high-technology spin-off firms identified in Catalonia by 1996 survey, but half of these set up since 1993 (Suris et al., 1998).

research institutes, and that there is a high rate, at least initially, of subsequent networking and collaborative activity between spin-offs and 'parents'. As stressed in chapter 2, the 1990s have witnessed dramatic technological changes and growth of new market opportunities for small, specialised technology-based start-ups in sectors such as information technology, computer software and internet applications, and biotechnology, leading to many new spin-offs. Such spin-offs have generally been localised as in the high-technology clusters studied by the network, while they are now often being encouraged by larger local parent firms[4] or universities. The spin-off process has also often played a key role in the growth of local 'mico-clusters' of firms in a particular sector, as for example with the development of the Cambridge region's striking micro-cluster of ink-jet printing firms using advanced micro-electronics technology, following the spin-off of the region's first successful such firm, Domino, from the R&D consultancy Cambridge Consultants in 1979. At the dawn of the twenty-first century, an active and often intense local spin-off process thus appears to be a hallmark of successful European regional high-technology clusters, leading to the localised diffusion between firms, and subsequent further development of, embodied expertise, innovative technologies, and managerial and research competences. The spin-off process thus stimulates new knowledge and shared learning: high spin-off regions are also high-learning regions.

8.4 Labour Market Recruitment and the 'Knowledge Carriers'

A second very important regional collective learning mechanism stressed in the theoretical discussion of section 8.1 is local labour market recruitment and movement between local firms, universities, research institutes and other organisations of highly-skilled employees, researchers, scientists, engineers and managers. As Lorenz (1996) stresses, 'mobile workers [are] the carriers of knowledge on the local labour market'. And the movement within a high-technology regional cluster of such workers, individually or in teams, represents a further important type of 'untraded interdependency' (Storper, 1995) between local firms, resulting in the transfer of 'embodied expertise', enhanced informal links, and a deepening and broadening of the regional pool of knowledge. Local universities, with their continuous output of young qualified scientists and engineers, may play a particularly significant role in this regard, with graduate and postgraduate recruitment by local firms helping local dissemination and commercial application of new scientific knowledge derived from university research. Certainly Longhi (1999, p. 336) argues that in the Sophia-Antipolis case, the creation for the first time in the 1990s of a genuinely 'innovative milieu' depended crucially on 'the emergence of a [highly-qualified] local labour market with the siting of the University of Nice in Sophia-Antipolis'.

That said, however, it is clear that most regional labour market recruitment of highly-qualified staff in the European high-technology clusters studied by the network involves flows of expertise between firms, both small and large. This can be illustrated by reference to the Cambridge high-technology cluster (Table 8.3). Notwithstanding the small size of the Cambridge region labour market, firms in the Cambridge region represent the second most frequent source of recruitment by high-technology SMEs[5] of both researchers and managers, out of the six categories listed. Recruitment from firms elsewhere in the UK (and overwhelmingly from adjacent southeast England) represents the most frequent source. In addition, one-fifth of Cambridge high-technology firms had recruited at least one recently-appointed research worker directly from Cambridge University. Such intra-regional recruitment diffuses technological and organisational knowledge, strengthens personal networks, and enables new combinations of knowledge to be assembled and deployed to develop new innovative products which may straddle traditional sectoral boundaries, as noted in chapter 2, section 2.6.

Detailed empirical evidence from other network case-study regions on the extent of intra-regional recruitment by technology-intensive SMEs of highly-qualified staff from other local firms and organisations is unfortunately more limited. The evidence collected for these other cases as well as Cambridge is summarised in Table 8.4.

In several of these regions, such as Cambridge, Oxford, Sophia-Antipolis, Pisa and Piacenza, the intensity of regional labour market recruitment of highly-skilled workers is noteworthy given the small size of the particular labour market involved. Not only does small regional size limit the potential for local recruitment, but highly-qualified and high-income scientists, engineers and other research staff, and managers experienced in guiding technology-based start-ups, are known to be exceptionally mobile geographically, usually operating within national if not international rather than local labour markets (Green and McKnight, 1996). Yet firms in these regions report relatively high levels of local recruitment of these key 'knowledge workers'. The network's research in fact suggests

Table 8.3 Research and managerial staff recruitment and mobility within the Cambridge region

Firms reporting recruitment of at least one of their last three research and management staff from:

	Research staff		Management staff	
	No.	%*	No.	%*
Cambridge University	7	19	2	6
Other Cambridge firms or organisations	13	35	12	39
Other UK universities	10	27	3	10
Other UK firms/organisations	15	41	18	58
Overseas universities	4	11	1	3
Overseas firms/organisations	3	8	7	23

* Percentages are of total respondents to this question (37 for research staff, 31 for management staff).

Source: Keeble et al, 1999.

Table 8.4 Regional labour market recruitment and flows of managerial and technological expertise

- **Göteborg** – exceptionally high level of recruitment by new technology-intensive SMEs of technical employees from within the Göteborg region: five of the six most frequently used and valued sources are local, not national or international: 'the most striking finding ... is the very high frequency of usage and the very high rating in importance of local universities'; the Göteborg study 'provides clear evidence of active processes of intra-regional movement of technical expertise' (Lindholm Dahlstrand, 1999, p. 386).

- **Cambridge** – 48% of firms report existing links with other local firms because of the movement of people between them, and 77% of these links are rated as important or very important for the firm's development: other firms in Cambridge region are now the second most frequent source of high-technology SME recruitment of researchers and managers (on average 38% of research staff recruitment, 33% of managerial staff recruitment, now from Cambridge region) (Keeble et al., 1999).

- **Oxford** – 58% of firms report links with other Oxford region firms because of movement of people between them, and 56% of these rated the

> links as important or very important for the firm's development: other Oxford region firms the second most frequent source of managerial recruitment by high-technology SMEs (other UK firms first), but researcher recruitment more frequent from UK firms and universities outside the region (Oxford University and Oxford firms third and fourth most frequent) (Lawton Smith, 1997a).
>
> - **Sophia Antipolis** – 'doctoral training in association with existing research institutes has helped create a labour market of highly-qualified workers and has changed the nature of the relations between firms and research: roughly 2,000 students and as many researchers are today working or studying in Sophia-Antipolis ... this local labour market is now of great significance in relation to activities linked to information technologies' (Longhi, 1999, p. 337).
>
> - **Grenoble** – 'the growth in importance of engineers, researchers and technicians in the Grenoble region has been central to the transformation of the regional economy. The development of skilled manpower in computing (software and hardware) ... has powerfully influenced firm location in more profound ways than just as a result of traditional labour market considerations ... localized collective learning in the Grenoble region is crucially dependent on the pool of local competencies' (de Bernardy, 1999, pp. 349, 351).
>
> - **Pisa, Piacenza, NE Milan** – small, highly innovative firms engaged in radical product innovation in these three milieux are distinctively characterised by high levels of recruitment of technicians and scientists from local firms and research centres, high local labour turnover, and high levels of firm spin-off: 'radical innovation activity by the smallest firms depends significantly on both [high] turnover of the labour force within the firm, and spin-off mechanisms: both describe mechanisms of tacit transfer of collective learning within the district' (Capello, 1999, p. 363).

that as these regional SME clusters have grown in size and achieved some degree of critical mass, so intra-regional flows of technological and managerial expertise have grown and intensified. Increased recent multinational firm R&D laboratory investment in several of these regions, as noted in chapter 2, is just one reflection of the significance of such flows, and the growing realisation by such large firms of the importance of access to the region's highly-qualified labour market, pool of tacit and embedded knowledge, and associated collective learning capability.

Two further important points warrant emphasis concerning the importance of the movement of skilled workers within the clusters studied and its significance for collective learning. First, it is clear that not just recruitment but also retention of such workers within the region is very important if cumulative build-up of collective knowledge and learning capacity is to be achieved. It is thus of central importance that virtually all of the clusters studied are distinctive in offering their skilled workers a high level of residential amenity, environmental attractiveness and quality of life, in terms of townscape, climate, culture, or recreational opportunities. The importance of this criterion of regional residential attractiveness for highly-mobile and 'choosy' knowledge workers and their families cannot be overstated. In the Cambridge case, for example, the region's provision of 'an attractive local living environment for staff and directors' was rated by the sample of high-technology SMEs surveyed as the single most important region-specific advantage for their firm's development out of 19 possible advantages, with 80 per cent of firms rating it as moderately, considerably or extremely important (Keeble et al., 1999, p. 325: see also Keeble, 1989). This distinctive attribute of the European 'high-learning' regional clusters studied is directly associated with these regions' ability to retain and grow the regional pool of skilled scientists, engineers, managers and entrepreneurs on which development of a collective learning capacity crucially depends.

Secondly, however, internal regional labour market recruitment and flows of skilled workers are insufficient on their own as a basis for cumulative knowledge development and technological innovation, given the much wider national and global contexts in which technological change and competition now operate. Regional collective learning via labour market processes thus also generally requires some inflow of expertise, knowhow and new embodied knowledge from other technologically-innovative regions and even countries. As de Bernardy (1999, p. 351) notes, 'localized collective learning in the Grenoble region is crucially dependent on the pool of local competencies, and ... local critical mass': but he also stresses that 'if, however, links with other milieux are weak, critical mass will be insufficient and dynamic entrepreneurs and researchers may move away from the area'. Successful high-technology clusters need to be linked into wider national and international labour markets, a characteristic clearly evident empirically from the network's case studies (see Table 8.3).

8.5 Inter-Firm Links, Networks and Collaboration in Regional Knowledge Development

Knowledge development through conscious interaction and collaboration between firms and other local organisations (firms, universities, research institutes), as theorised by workers such as Camagni and Lorenz, represents the most intuitively-obvious regionally-based process of collective learning in a high-technology cluster. Though conscious, such networking is often informal rather than formal, involving both strong and weak ties (Yeung, 1994) and the development of a local 'industrial atmosphere' of personal relationships and trust. It is important to note, as Storper (1995) points out (see also chapter 1, section 1.7), that traditional input-output (supplier-customer) traded linkages are often relatively limited in contemporary regional high-technology clusters, a finding corroborated by the network's research (with the interesting exception of the Italian manufacturing-based high-technology regions, which do exhibit strong local supplier-customer links). The networks identified by the present research more commonly involve 'untraded interdependencies', people-based informal links, and knowledge transfer and sharing, rather than movement of goods or components: in Grenoble, 'learning by imitation and adaptation ... has mainly operated through informal local networking linked to entrepreneurs' address books, word of mouth contacts and webs of personal relationships built up since university education and through common social activities' (de Bernardy, 1999, p. 350).

Nonetheless, as Hughes' (1998) research in Britain has clearly demonstrated (see chapter 2, section 2.7), high-technology SMEs are in general markedly differentiated from 'conventional' SMEs in exhibiting exceptionally high levels of inter-firm and organisation networking and collaborative activity. This functional characteristic has clear spatial expression and identity, in that the network's research reveals that a characteristic hallmark of a 'high-learning' technology-based SME regional cluster is indeed a high level of local networking, collaboration and linkage. Detailed evidence is provided in Table 8.5.

The network's research, as summarised in Table 8.5, clearly reveals that Europe's successful technology-based clusters are invariably characterised by active and relatively intense local networking, involving the exchange and development of new knowledge and enhanced collective learning, between local firms and organisations, notwithstanding these clusters' different origins and structural characteristics (see Chapter 2, section 2.5). Network-

Collective Learning Processes in European High-Technology Milieux 215

Table 8.5 Inter-firm links and networks in regional knowledge sharing and collective learning

- **Cambridge** – 76% of high-technology SMEs report 'close links' with other local firms, most frequent links being with suppliers and subcontractors (68% have 'important' local links), service firms (53%), customers (32%) and research collaborators (29%): 50% of SMEs report research links with Cambridge University since formation, 56% of which regard these links as 'important' for the firm's development (Lawson, 1997b).

- **Oxford** – 46% of high-technology SMEs report 'close links' with other local firms, and 87% of these rate such links as important or very important to them: 54% report links with Oxford University or local public research institutes (66% rated these as 'important' for firm's development): large indigenous technology-based firm, Oxford Instruments, appears to have played a valuable role in fostering local high-technology SME subcontracting links (Lawton Smith, 1997b).

- **Grenoble** – 'Grenoble has developed over recent decades a dynamic innovation system, providing local SMEs with a range of networking relationships' ... 'the number of research contracts between Grenoble firms and university research institutes has grown substantially in recent years': local networks are especially important for new start-ups/spin-offs, while mature SMEs network selectively both locally and nationally: growth in 'critical mass' very important in creating opportunities for local selective networking (de Bernardy, 1999).

- **Sophia-Antipolis** – 'the development of Sophia-Antipolis ... was characterized until recently by the weakness of local interactions ... [reflecting] the location of a disparate and largely unrelated geographical cluster of firms' (Longhi, 1999, p. 335): in the 1990s, however, growth of numerous 'small and micro firms, with strong links with research and close interrelationships' in the information technology sector; 'this cluster is ... embedded within numerous different local formal and informal networks ... linkages and interactions between these firms are very important and involve personal contacts, exchange of information, even sharing of some niches in terms of micro-specialization' (Longhi, 1999, p. 341).

- **Göteborg** – links and networks with 'parent' organisations are very important for spin-offs and new start-ups (73% of technology-intensive SMEs report such links): other important links for 'competence

> development' at start-up are with Göteborg universities and Swedish customers elsewhere (both 21% of firms): after 10 years, however, local links decline, wider national Swedish links increase (Lindholm Dahlstrand, 1999).
>
> - **Munich** – 'Munich's innovative environment' is characterised by 'intensive intra-regional co-operative networking': 42% of R&D-intensive SMEs (manufacturing and services) co-operate for innovation with regional partners, with average numbers of regional partners per R&D-intensive firm of 4.5 manufacturing firms, 3.0 service firms, and 2.4 research/education institutions: significant role of Siemens in regional networking (one-third of R&D-intensive SMEs rate innovation links with Siemens as 'significant' or 'very significant' for their business) (Sternberg and Tamásy, 1999).
>
> - **Pisa, Piacenza, NE Milan** – high-technology SMEs in these milieux reveal 'a clear cut propensity to develop relationships with local customers and suppliers' (81% report local customer linkages, 84% report local supplier linkages): local customer linkages involve 'a strong interchange of expertise and relative skills': there has been 'a marked tendency to strengthen the [customer] relationships ... over the last five years' (Campoccia, 1997).

ing and associated new knowledge development appear to be especially important for small, new and innovative firms. However, this review also suggests three important qualifications to this general conclusion.

First, there is considerable evidence that development of successful local networks and linkages, and of a consequent collective learning capacity, is time- and 'critical mass'-dependent. Younger and newer innovative milieux, such as the Oxford region compared with the more developed Cambridge case, exhibit fewer local linkages (although their importance to the fewer firms involved is greater), while in Sophia-Antipolis, the growth of active local networking is very recent and largely confined to the rapidly-burgeoning information technology small firm cluster. The Grenoble case study emphasises the critical mass issue particularly strongly, de Bernardy (1999, p. 351) arguing that 'critical mass in a specific technological field is becoming a necessary condition for sustaining new innovative activities' because of the growing need for specialisation in inter-firm networking, partnership selection, and collective learning. The clear implication for policy here is that the fostering of a successful learning-based regional cluster of high-technology SMEs has to involve long term support and policy persistence over several decades, even under favourable external and environmental circumstances.

A second interesting issue concerns the role of large technology-based firms in networking, collaboration and collective learning in these European SME clusters. Clusters which contain locally-headquartered firms of this kind do appear to benefit rather than suffer from their activities, through their role in orchestrating local networks of SMEs as subcontractors and suppliers. This appears to be the case with Siemens in Munich and Oxford Instruments in Oxford. Sternberg and Tamásy (1999, p. 374) report, for example, not only that one-third of all Munich high-technology SMEs have significant innovation-related links with Siemens, but that these SMEs exhibit significantly stronger growth in turnover and employment than SMEs which do not possess such links. They argue that 'the Granovetterian characteristics of embeddedness, trust as the basis of co-operation in innovation, and the more informal basis can be clearly empirically proven – a fact that applies to contacts between R&D-intensive SMEs and Siemens, as well as to other partners in the region'. An even more interesting finding, this time from the Cambridge study, is that externally-based large firms may also enhance local networking and collective learning, through the activities of their local subsidiaries. In this case, and contrary to the simple hypothesis that subsidiaries would be less likely to engage in local networking and partnerships because of easy access to external company-based knowledge networks and resources, 'subsidiaries consistently reported higher numbers of all kinds of inter-firm links' within the region than independent SMEs (Lawson, 1997b, p. 44). The difference was particularly marked for research links, while subsidiaries also valued their local links more highly (with the exception of customers). Local subsidiaries of large firms thus appear to be more, not less, embedded in regional networks and hence involved in local knowledge development and collective learning than other firms, again supporting the thesis that large firms are more likely to enhance than diminish the growth of regional clusters of technology-based SMEs.

The third qualification is that, as argued in chapter 1 (section 1.8), successful European high-technology SMEs need access to national and global innovation networks as well as to local network-based collective learning processes if they are to sustain longterm innovativeness and competitiveness in a globalising world. Notwithstanding the great importance of regional links and networks, such firms are almost invariably much more oriented to national and international markets than conventional SMEs (Keeble, 1994, pp. 211–12: Keeble et al., 1998: see also Table 2.1), and interact more with external innovation partners than with regional partners. Thus even in the Munich region (Sternberg and Tamásy, 1999), 58 per cent of such partner firms and organisations are external (20 per cent foreign), while in Cambridge, 'national and global innovation networks are ... appreciably more frequently rated as important than are local networks' (Keeble et al., 1999, p. 327). This is equally true in the Oxford case (Lawton Smith, 1997b, p. 63). High-learning regions thus are and need to be embedded not just in local, but in global and national, learning networks if they are to sustain competitiveness over time.

That said, however, the network's original research on the Cambridge case (Keeble et al., 1998, pp. 337–8) also breaks new ground in showing that the local and the global are not alternative, substitutable, networking environments for regionally-clustered high-technology SMEs. Far from outgrowing local networks, interdependencies and collective learning mechanisms, those technology-intensive SMEs which successfully develop global markets and research partnerships actually tend to exhibit greater local networking intensities than their less internationally-oriented counterparts. This is especially true for local research collaboration with both other firms and universities, with its direct implications for regional collective learning, as well as for local links with competitors and firms in the same sector (Keeble et al., 1998, p. 337). This striking finding suggests that regional embeddedness within a vibrant and networked learning-oriented milieu may actually enhance the likelihood of high-technology SMEs achieving success globally. Local and global innovation networks thus appear to be of simultaneous – and probably complementary – importance for the competitive success and growth of regionally-clustered technology-based SMEs.

8.6 Collective Learning and Innovation

As noted earlier in section 8.1, the existence of active regional collective learning processes and of a 'high-learning' regional milieu is likely to be evident in a high regional rate of firm product and service innovation, as a key output measure of the effectiveness of local learning processes. On this criterion, virtually all the European high-technology SME clusters studied appear to be characterised by active and effective learning mechanisms, in that their technology-based SMEs exhibit high rates of innovation in new products and services.[6] Evidence clearly substantiating this at a macro level is provided by the OST research by Barré, Laville and Zitt (1998) already described and mapped in Chapter 2 (section 2.5, Figure 2.2). This shows that on the basis of key science and technology indicators (density of technology patents and scientific publications relative to regional workforce in the 1990s), all the study regions are amongst the most highly technologically innovative (type 1 out of eight categories) in the European Union, with the exception of Göteborg (type 2) and Barcelona (type 3).

Although data is unfortunately more limited, the network's own micro-level research also reveals very high rates of product and service innovation by technology-based SMEs in those study regions for which survey results are available. Campoccia's research (1997, p. 26) into the Pisa, Piacenza and NE Milan high-technology clusters thus found that *all* the firms surveyed had developed new product innovations during the preceding five years, with 60 per cent claiming that these represented radical 'breakthrough' innovations, rather than just 'upgrading' innovations. Equally, both the Cambridge and Oxford regional clusters are characterised by very high rates of product and technological innovation. No less than 92 per cent of the high-technology SMEs surveyed in these two milieux[7] reported developing product or service innovations during the previous three years, with 65 per cent of firms deriving 50 per cent or more of current sales from these innovations. Some 69 per cent also rated their innovations as very or extremely original (4 or 5 on a scale of 1–5, from minimal to maximal originality). Finally, research on high-technology manufacturing and service SMEs in the Munich region reveals that no less than 121 out of 150 sampled R&D-intensive enterprises (81 per cent) had introduced new product innovations during the previous four years (1993–97). Amongst the subsample of manufacturing SMEs, product innovation rates were no less than 94 per cent.[8]

All three of these high-technology regional clusters thus exhibit exceptionally high levels of product innovation, compared with innovation rates reported by both conventional and high-technology SMEs outside such clusters. An authoritative measure of the latter is provided by the 1997 British national survey of 2,386 manufacturing and professional and business service SMEs conducted by Cambridge University's ESRC Centre for Business Research, which records product innovation rates over the preceding three years for both conventional (55 per cent) and high-technology (72 per cent) firms which are appreciably lower than those found in the high-technology clusters discussed above (Hughes and Moore, 1998, p. 94).[9] Again, this provides strong support for the contention that active collective learning processes in Europe's technology-based clusters powerfully enhance the innovativeness of locally-based SMEs. Detailed empirical investigation thus shows that high-technology clustering is strongly associated with both collective learning and innovativeness: high learning regions are also and invariably highly innovative regions.

8.7 Collective Learning, Knowledge Institutions, and New Regional Collective Initiatives

The network's research thus provides rich evidence of the existence in all the successful high-technology SME regional clusters studied of active local inter-firm and firm-organisation processes which promote learning, knowledge development, and exceptional levels of technological and product innovation. The key regional collective learning processes identified by this research – new firm spin-off and entrepreneurship, labour market recruitment, and networking, collaboration and linkage – arguably lie at the heart of the recent evolution and competitive success of the successful regional clusters studied. These processes operate primarily between individual firms, both small and large. But they are clearly also influenced and shaped by regional non-firm institutions and institutional support mechanisms. As a final observation from our comparative analysis of European high-technology regional clusters in the 1990s, then, the particular and potential role of two types of non-firm institution in the development of a regional collective learning capability perhaps warrants special emphasis.

8.7.1 Knowledge Institutions

The first of these non-firm institutions, already of course discussed in some detail in chapter 4, are knowledge institutions and especially major regional universities. As chapter 4 documents, the network's research indicates that in many cases, regional universities are playing a important contemporary role in promoting collective learning within their local high-technology SME cluster. This may take one or more of at least five different forms, namely:

- creating preconditions for regional collective learning;

Collective Learning Processes in European High-Technology Milieux 221

- generating local technology-based spin-offs;

- training scientists, engineers, researchers and other graduates;

- collaborating with local firms in research and technology development;

- establishing university-linked science parks and incubator centres.

Specific empirical evidence of the role which particular local universities play in relation to the three key regional collective learning processes of entrepreneur spin-off, SME research collaboration, and research and technical staff recruitment, is provided in Table 8.6. This illustrates the importance which local knowledge institutions can have in promoting the development of a locally-embedded collective learning capacity in a high-technology SME cluster.

Four brief general conclusions may perhaps be drawn from the network's research on the role of knowledge institutions specifically in relation to the development of active processes of regional collective learning and technology development. First, regional universities can play an important if not central role in establishing preconditions for learning in terms of informal regional networks of former students and researchers, and SME research cultures of collaborative innovation. This appears to be true in such cases as Cambridge, Oxford,[10] Göteborg, Grenoble and, perhaps, Helsinki. Secondly, university spin-offs can be an important, though secondary, source of new innovative firms and regional technology competences. These may even over time help re-orientate a region's economic structure away from traditional but declining sectors, as perhaps in the Göteborg case (Lindholm Dahlstrand, 1998, p. 4).[11] Thirdly, formal research links between SMEs and local universities are probably less important for collective learning than is often assumed: a minority of R&D-intensive SMEs do collaborate with local universities, but university links frequently reflect the need for access to specialised technological expertise and are therefore more often national and global than regional. Fourth, recruitment of highly-qualified science, engineering and management graduates from local technology-focussed or major scientific universities appears to be of growing importance for high-technology SMEs in most of the regions studied. This is one of the most crucial ways in which local knowledge institutions can help shape and foster the growth of a territorially-embedded technology cluster, provided that the region's

222 *High Technology Clusters, Networking and Collective Learning in Europe*

Table 8.6   Knowledge centres and regional collective learning: European examples

> - **Spin-offs**
>   17% of Cambridge high-technology SMEs from Cambridge University:
>   18% of Oxford high-technology SMEs from Oxford University:
>   240 direct spin-offs from Chalmers University of Technology, Göteborg, 1960–93 (87% still operating):
>   rapid recent 1990s growth in spin-offs from Sophia-Antipolis government research institutes (INRIA).
>
> - **Research collaboration**
>   50% of Cambridge firms report research links with Cambridge University since formation: but only 56% of these rate the links as important for development: universities elsewhere (UK, globally) more important:
>   Oxford high-technology SMEs rate Oxford University research links more highly (44% of firms with links rate them very or extremely important):
>   Research links with local universities the second most important source of competence development (after customers) at start-up for Göteborg new high-technology firms: but these diminish with time:
>   but university research links relatively unimportant in Munich region (regional cluster dominated by manufacturing, strong SME links with Siemens).
>
> - **Research and managerial staff recruitment**
>   53% of Cambridge firms have at least one researcher with Cambridge degree (19% of recent research staff recruits from Cambridge university):
>   Göteborg universities the most important single source of technical employees for region's new high-technology SMEs (used by 58% of firms):
>   University of Nice-Sophia-Antipolis now a major source of engineers, computer scientists, management graduates for local firms
>   31% of graduate engineers from Grenoble's INPG engineering institute recruited by local firms in Rhône-Alpes (second only to Paris).

residential amenities and environment are sufficiently attractive to retain such mobile and locationally-selective individuals.

8.7.2 New Regional Collective Initiatives

A second type of non-firm institutional influence on regional collective learning processes which warrants specific reference is the recent rise of new regional coalitions and collective initiatives. In several of our high-technology SME clusters, new regional collective initiatives have emerged in the late 1990s aimed directly at encouraging processes of regional collective learning. This very interesting and novel trend involves regional coalitions involving some or all of the following organisations: SMEs, large firms, universities, public research institutes, local government, trades unions, and business support and training agencies. Their aims include strengthening local business support structures, overcoming constraints on business growth in the region by, for example, improving access to capital including venture capital, fostering collaborative activity between local firms and universities, and marketing the region globally via Internet and other publicity. Their development appears to reflect growing awareness both of intensifying global competition, and of emerging constraints on the continuing growth of these clusters arising directly from successful previous expansion. The latter include opposition to further business expansion from environmental groups, increasing planning constraints, and infrastructure bottlenecks. Collective action is now seen by business representatives, politicians and planners in these regions as essential if constraints are to be resolved, and the region's prosperity sustained and enhanced. A key element in these initiatives is increased networking and collective learning between local firms and institutions, to share best practice business knowledge, improve training provision, and strengthen further collaborative links with local knowledge institutions. Examples of such collective initiatives have emerged in Cambridge, Sophia-Antipolis and Grenoble. Table 8.7 provides detailed information on the Cambridge case.

The simultaneous but independent emergence in different European high-technology regional clusters of such regional collective initiatives as an indigenous response to perceived problems, needs and increasing global competition is noteworthy and potentially very significant for strengthening the regional institutional framework within which collective learning can flourish.

8.8 Conclusions

This chapter has attempted to review and synthesise the detailed empirical evidence generated by the European network's research concerning the nature and extent of regional collective learning mechanisms or processes operating in the European high-technology SME clusters studied. Following

Table 8.7   New regional collective initiatives in the Cambridge region, 1997-99

Cambridge Network High-technology business-led initiative (Herman Hauser, David Cleevely) to raise global profile of and increase local networking by Cambridge IT companies. Incorporates Cambridge Connect, an interactive website modelled on San Diego Connect relating to the Cambridge subregion and business support facilities available to other businesses.
Greater Cambridge Partnership Established 1998 to develop consensus between local business, government (county and districts) and university on future economic strategy for Cambridge region, in face of constraints and conflicts: strong business involvement, working groups investigating business development, planning and capacity constraints, employee skills, investment promotion and attraction.
Cambridge Futures Academic and business alliance investigating, with private sector funding, alternative 50-year scenarios for accommodating anticipated growth: report published May 1999.
New Science Parks Cambridge region now contains five Science Parks (Cambridge Science Park, St John's Innovation Park, Melbourn Science Park, and two new parks, Granta Park and Cambridge Research Park) plus approved plans for a substantial biotechnology incubator at Hinxton Hall (associated with the Wellcome Trust's Sanger Centre Human Genome Project), a further small biotechnology park at the Babraham Institute, a new Bioscience Innovation Centre on St John's Innovation Park, and a doubling of the size of the Scientific Generics R&D centre at Harston.
St John's Innovation Centre as Regional High-Technology Business Support Agency Established 1987 to house new high-technology start-ups unable to afford commercial rents: 60 current firms, plus 80 more 'graduates' who have moved to larger premises: since 1994, growing regional role through Financial Packaging Service (advice on financial and business planning: now 20-25 new clients every month), Anglia Enterprise Network (experts – 'ferrets' – to identify new business opportunities emerging from Cambridge University science departments), Shell Technology Enterprise Programme (student

placements with local high-technology firms), Resource Centre for local high-technology firms (in collaboration with the Cambridge University Judge Institute of Management Studies, and Institute of Manufacturing), European Relay Centre, High-Technology Business Club, Business Angel Database ...

Proliferating Venture Capital Funds
Prelude Technology Investments
Amadeus Capital (includes Microsoft venture capital fund)
Cambridge Research and Innovation (CRIL)
Cambridge Quantum Fund
Gateway Fund
3i plc Cambridge Office

New Cambridge-based Eastern Region Initiatives
Two major new regional development organisations, the East of England Investment Agency and the East of England Development Agency, have both established their headquarters in Cambridge since 1997, following the Government Office for the Eastern Region in 1994.

Source: modified and updated from Garnsey, 1998.

some consideration of appropriate theory, it has argued that such processes require the previous development of appropriate preconditions for learning, in terms of common regional culturally-based rules and codes of inter-firm behaviour, collaboration and interaction, and that in most of the regions studied, these preconditions have been established, albeit over a considerable period of time and in very different regional contexts, through the actions of key regional 'collective agents' such as universities or influential large firms. It has also documented, with a wealth of empirical detail, the great importance of three key collective learning processes in the growth and innovative activity of these clusters, namely new firm and entrepreneur spin-offs, labour market flows of highly-qualified staff, and inter-firm networking, collaboration and linkage. Regions characterised by high intensities of these three processes can be categorised as 'high learning regions', possessing an in-built capacity for continuing and successful development of commercially-valuable new knowledge, products and services. Finally, it has shown conclusively that high-technology SMEs in these regions are, arguably as a result, exceptionally innovative, and has stressed the particular role of knowledge institutions and new regional collective initiatives in encouraging and sustaining regional collective learning capacity. All these findings have implications for regional and

226 High Technology Clusters, Networking and Collective Learning in Europe

national government policies, and it is to consider these that the final chapter of this book now turns.

Notes

1. Whittaker's recent (1999, p. 11) work on high-technology entrepreneurs in Britain reveals that over two-thirds (68.6%) of a sample of 510 technology-based single-site independent enterprises employing less than 200 workers (92% less than 100 employees) were established by two or more founders, rather than by a single individual.
2. Again, Whittaker (1999, p. 10) found that 40 per cent of the entrepreneurs surveyed actually received help from their parent organisation in setting up their new enterprise.
3. See note 1 above.
4. See note 2 above. 'The size of [the] previous business is particularly interesting. For almost half (45.7%) of the participants, the business had 300 or more employees, and for 60.3%, 100 or more employees. For only a quarter (26.3%) the business had less than 50 employees. The notion that small firms are better incubators than large firms ... might apply to 'traditional' small firms, but not ... it would seem, the high tech sector' (Whittaker, 1999, p. 7).
5. Table 8.3 reports results from a 1996 survey of a stratified random sample of 50 Cambridge region technology-based firms, carried out by the ESRC Centre for Business Research of Cambridge University: see Keeble et al. (1999).
6. The only exception here is probably the Barcelona region, with as yet only a small and disparate cluster of technology-intensive SMEs.
7. These are hitherto unpublished figures, based on combined responses from 100 technology-based SMEs sampled equally from the two regions.
8. I am indebted to Gero Stenke, Christine Tamásy and Rolf Sternberg for this unpublished information, derived from their 1998 SME survey in the Munich region.
9. This survey involved a stratified random sample of SMEs from throughout Great Britain. High-technology SMEs were defined using the sectorally-based Butchart classification set out in chapter 1, Table 1.1. Note that product innovation rates were very similar both for high-technology manufacturing (72%) and service (73%) SMEs, and for conventional manufacturing (59%) and service (50%) SMEs.
10. But note that Cambridge and Oxford Universities have had different explicit policies towards local technology transfer and the intellectual property involved in spin-offs, with the former adopting a more laissez-faire, the latter a more restrictive, approach (Garnsey and Lawton Smith, 1998, pp. 446–7).
11. 'NTBFs in West Sweden might be in the process of deepening its technology base and enlarging the science base' (Lindholm Dahlstrand, 1998, p. 4).

References

Aydalot, P. (ed.), (1986), *Milieux Innovateurs en Europe*, GREMI, Paris.
Aydalot, P. and Keeble, D. (eds) (1988), *High Technology Industry and Innovative Environments: the European Experience*, Routledge, London.

Barré, R., Laville, F. and Zitt, M. (1998), *The Dynamics of S&T Activities in the EU Regions*, Observatoire des Sciences et des Techniques (OST), Paris.

Camagni, R. (1991a), '"Local 'Milieu", Uncertainty and Innovation Networks: Towards a New Dynamic Theory of Economic Space', in R. Camagni (ed.), *Innovation Networks: Spatial Perspectives*, Belhaven Press, London, pp. 121–42.

Camagni, R. (ed.), (1991b), *Innovation Networks: Spatial Perspectives*, Belhaven Press, London.

Camagni, R. (1995), 'High-Technology Milieux in Italy and New Reflections about the Concept of "Milieu Innovateur"', paper presented at European Workshop on High-Technology Enterprise and Innovative Regional Milieux, Cambridge, 3–4 March.

Campoccia, A. (1997), 'Inter-SME Linkages in Italian High-Technology Milieux: Empirical Evidence', in D. Keeble and C. Lawson (eds), *Networks, Links and Large Firm Impacts on the Evolution of Regional Clusters of High-Technology SMEs in Europe*, ESRC Centre for Business Research, University of Cambridge, Cambridge, pp. 25-39.

Capello, R. (1998), 'Collective Learning and the Spatial Transfer of Knowledge: Innovation Processes in Italian High-Tech Milieux', in D. Keeble and C. Lawson (eds), *Collective Learning Processes and Knowledge Development in the Evolution of Regional Clusters of High-Technology SMEs in Europe*, ESRC Centre for Business Research, University of Cambridge, Cambridge, pp. 19–37.

Capello, R. (1999), 'Spatial Transfer of Knowledge in High Technology Milieux: Learning Versus Collective Learning Processes', *Regional Studies*, vol. 33, pp. 353–65.

Chapman, J. (1999), 'The Start-Ups: "Magic Slate" Could be Worth Millions', *Cambridge Evening News, Business and Finance Supplement*, 27 July, p. 9.

de Bernardy, M. (1999), 'Reactive and Proactive Local Territory: Co-operation and Community in Grenoble', *Regional Studies*, vol. 33, pp. 343–52.

Garnsey, E. (1998), 'Dynamics of the Innovative Milieu: Examples from Cambridge', in D. Keeble and C. Lawson (eds), *Collective Learning Processes and Knowledge Development in the Evolution of Regional Clusters of High-Technology SMEs in Europe*, ESRC Centre for Business Research, University of Cambridge, Cambridge, pp. 117–21.

Garnsey, E. and Lawton Smith, H. (1998), 'Proximity and Complexity in the Emergence of High Technology Industry: The Oxbridge Comparison', *Geoforum*, vol. 29, pp. 433–50.

Green, A. and McKnight, A. (1996), 'The Changing Distribution and Composition of the Highly Qualified', unpublished conference paper, Institute for Employment Research, University of Warwick.

Henry, N. and Pinch, S. (2000), 'Spatialising Knowledge: Placing the Knowledge Community of Motor Sport Valley', *Geoforum*, forthcoming.

Hughes, A. and Moore, B. (1998), 'High-Tech Firms: Market Position, Innovative Performance and Access to Finance', in A. Cosh and A. Hughes (eds), *Enterprise Britain: Growth, Innovation and Public Policy in the Small and Medium Sized Enterprise Sector 1994–1997*, ESRC Centre for Business Research, University of Cambridge, Cambridge, pp. 86–98.

Kauranen, I. (1997), 'The Institutional Framework for Research and Technology Development Intensive Small and Medium Sized Enterprise Developments in the Helsinki Region', in D. Keeble and C. Lawson (eds), *University Research Links and Spin-Offs in the Evolution of Regional Clusters of High-Technology SMEs in Europe*, ESRC Centre for Business Research, University of Cambridge, Cambridge, pp. 2–12.

228 High Technology Clusters, Networking and Collective Learning in Europe

Keeble, D. (1994), 'Regional Influences and Policy in New Technology-Based Firm Creation and Growth', in R. Oakey (ed.), *New Technology-Based Firms in the 1990s*, Paul Chapman, London, pp. 204–18.

Keeble, D. (1989), 'High-Technology Industry and Regional Development in Britain: the Case of the Cambridge Phenomenon', *Environment and Planning C: Government and Policy*, vol. 7, pp. 153–72.

Keeble D., Lawson C., Lawton Smith H., Moore B. and Wilkinson F. (1998), 'Internationalisation Processes, Networking and Local Embeddedness in Technology-Intensive Small Firms', *Small Business Economics*, vol 11, pp. 327–42.

Keeble D., Lawson C., Moore B. and Wilkinson F. (1999), 'Collective Learning Processes, Networking and 'Institutional Thickness' in the Cambridge Region', *Regional Studies*, vol. 33, pp. 319–32.

Lawson, C. (1997a), 'Territorial Clustering and High-Technology Innovation: from Industrial Districts to Innovative Milieux', *ESRC Centre for Business Research, University of Cambridge, Working Paper* 54.

Lawson, C. (1997b), 'Local Inter-Firm Networking by High-Technology Firms in the Cambridge Region', in D. Keeble and C. Lawson (eds), *Networks, Links and Large Firm Impacts on the Evolution of Regional Clusters of High-Technology SMEs in Europe*, ESRC Centre for Business Research, University of Cambridge, Cambridge, pp. 41–55.

Lawson, C. (1999), 'Towards a Competence Theory of the Region', *Cambridge Journal of Economics*, vol. 23, pp. 151–66.

Lawson, C. and Lorenz, E. (1999), 'Collective Learning, Tacit Knowledge and Regional Innovative Capacity', *Regional Studies*, vol. 33, pp. 305–17.

Lawton Smith, H. (1997a), 'University and Public Sector Research Laboratory Links and Technology Transfer in the Oxford Region', in D. Keeble and C. Lawson (eds), *University Research Links and Spin-Offs in the Evolution of Regional Clusters of High-Technology SMEs in Europe*, ESRC Centre for Business Research, University of Cambridge, Cambridge, pp. 24–38.

Lawton Smith, H. (1997b), 'Inter-Firm Networks in Oxfordshire', in D. Keeble and C. Lawson (eds), *Networks, Links and Large Firm Impacts on the Evolution of Regional Clusters of High-Technology SMEs in Europe*, ESRC Centre for Business Research, University of Cambridge, Cambridge, pp. 57–66.

Lazaric, N. and Lorenz, E. (1997), 'Trust and Organisational Learning During Inter-firm cooperation', in N. Lazaric and E. Lorenz (eds), *The Economics of Trust and Learning*, Edward Elgar, London.

Lindholm Dahlstrand, A. (1998), 'The Development of Technology-based SMEs in the Goteborg Region', in D. Keeble and C. Lawson (eds), *Regional Reports*, ESRC Centre for Business Research, University of Cambridge, Cambridge.

Lindholm Dahlstrand, A. (1999), 'Technology-based SMEs in the Göteborg Region: Their Origin and Interaction with Universities and Large Firms', *Regional Studies*, vol. 33, pp. 379–89.

Lorenz, E. H. (1992), 'Trust, Community and Co-operation: Toward a Theory of Industrial Districts', in M. Storper and A.J. Scott (eds), *Pathways to Industrialisation and Regional Development*, Routledge, London.

Lorenz, E. (1996), 'Collective Learning Processes and the Regional Labour Market', unpublished research note, European Network on Networks, Collective Learning and RTD in Regionally-Clustered High-Technology SMEs.

March, J.G. (1991), 'Exploration and Exploitation in Organisational Learning', *Organisation*

Science, vol. 2, pp. 71–87.
Maskell, P. and Malmberg, A. (1999), 'Localised Learning and Industrial Competitiveness', *Cambridge Journal of Economics*, vol. 23, pp. 167–85.
Ratti, R., Bramanti, A. and Gordon, R. (eds) (1997), *The Dynamics of Innovative Regions: The GREMI Approach*, Ashgate, Aldershot.
Scott, A. (1988), *New Industrial Spaces: Flexible Production Organisation and Regional Development in North America and Western Europe*, Pion, London.
Segal Quince Wicksteed (1985), *The Cambridge Phenomenon: The Growth of High Technology Industry in a University Town*, SQW, Swavesey, Cambridge.
Sternberg, R. and Tamásy, C. (1999), 'Munich as Germany's No. 1 High-Technology Region: Empirical Evidence, Theoretical Explanations and the Role of Small Firm/Large Firm Relationships', *Regional Studies*, vol. 33, pp. 367–77.
Storper, M. (1995), 'The Resurgence of Regional Economies, Ten Years Later: The Region as a Nexus of Untraded Interdependencies', *European Urban and Regional Studies*, vol. 2, pp. 191–221.
Storper, M. (1997), *The Regional World*, Guilford Press, New York.
Suris, J.M., Escorsa, P., Guallarte, C. and Maspons, R. (1998), 'The Creation of University Spin-Off Companies in Catalonia', unpublished research paper, Enterprise and Science Foundation, Autonomous University of Barcelona, Barcelona.
Whittaker, H. (1999), 'Entrepreneurs as Co-operative Capitalists: High Tech CEOs in the UK', *ESRC Centre for Business Research, University of Cambridge Working Paper* 125.
Yeung, H.W. (1994), 'Critical Reviews of Geographical Perspectives on Business Organisations and the Organization of Production: Towards a Network Approach', *Progress in Human Geography*, vol. 18, pp. 460–90.

9 Concluding Reflections: Some Policy Implications

FRANK WILKINSON AND BARRY MOORE

9.1 Introduction

An important part of the TSER European network's activities has been to explore the policy implications emerging from the different themes that have formed the focus of the collaborative discussions and presentations of regional case studies. A number of the findings reported earlier emerge as being of particular relevance for policy. Firstly, the case studies reveal that clusters of high-technology SMEs play an important role in strengthening innovative capability and competitiveness at a regional/local, national and EU level. Secondly, the case studies identify key processes of collective learning operating within local innovative milieux which provide powerful mechanisms for supporting innovative activity in SMEs and the diffusion of innovation and knowledge. These mechanisms include academic and company spin-offs, the mobility of skilled labour and a variety of formal and informal inter-firm links. The importance of effective university/research institute-industry relationships in strengthening the innovative potential of the cluster has particular implications for both technology transfer policies and local and regional economic development. Thirdly, the case studies confirm an important positive contribution from different local/regional institutions in facilitating local networking and providing channels for the distribution of knowledge and information in the local and regional productive system. This has implications for policy support for the institutional infrastructure at the local and regional level. Fourthly, the case studies demonstrate the importance of regional clusters of high-technology SMEs for the contribution they make to regional/local economic development and restructuring. Finally, the case study evidence suggests that dynamic localised clusters of high-technology SMEs, universities and research institutes can intensify existing economic and social disparities between regions and countries. Typically, the high-technology clusters are located in the relatively high growth regions and localities and it is clear from the different country experiences that not all regions provide the conditions necessary for the growth and development of such clusters.

The central focus of this final chapter is on the implications of the case studies for innovation and competitiveness policies at national and regional levels, although it should be recognised that there are important implications for policies targeted towards supporting small high-technology firms, and those concerned with reducing regional economic and social disparities.

9.2 The Theoretical Rationale for Policy

The rationale for policy intervention and the design or form that intervention takes turns critically on the theoretical framework within which the case studies are considered and analysed. From the traditional neo-classical perspective the focus and rationale for policy intervention is 'market failure'. The supply of knowledge is impeded because it readily becomes a public good so that the returns from its production are difficult to appropriate, because of uncertainty about its commercial benefits, especially at the level of basic research, and because of significant indivisibilities and economies of scale in its production. Uncertainty of outcomes and the necessary scale of expenditure increases the risk of under-investment by the market in knowledge production justifying state intervention by, for example, the funding of education, training and basic research. Policy informed by neo-classical theory therefore tends to be dominated by a linear model of technical development. Public investment in basic science and in education and training is designed to overcome the problems of uncertainty and to secure scale economies. This produces a flow of knowledge that feeds innovation undertaken later by business.

The closer knowledge is to commercial exploitation, however, the higher is the risk of spill-over and the greater the threat to the profitability of innovation. In response to this, neo-classical theorists have embraced the Schumpeterian notion of the importance of restrictions on competition for encouraging innovators, although not its more dynamic implications. They accept that patenting, monopoly control, inter-firm collaboration and other forms of restricting access to new ideas counters the risk of under-investment in innovation by allowing the more effective appropriation of technology rents and thereby increasing the incentive to develop new products and processes. Here, theoretically at least, the welfare loss of 'market failure' resulting from restrictions on competition are seen as being offset by the gains from more advanced technology. From this perspective, the problem for policy and its implementation is how to evaluate these costs and benefits in particular cases.

An alternative, and more dynamic, rationale for high technology policy is offered by evolutionary economics and competence theories of the firm and region.[1] These and related theories have provided the theoretical framework for the collaborative research of the TSER European network. The key concept is competences, which are seen as determining the technical, marketing, managerial and other capabilities of firms and regions and therefore their competitive performance. The basis of competences is the shared knowledge organisations hold which is embodied in the routines and procedures which co-ordinate the joint activities of members and enable them effectively to communicate and work with each other. The 'dynamic' capabilities of organisations are their ability to raise business performance by improving their competences and developing new ones by incorporating new information into the knowledge base. These learning processes within knowledge systems are a central focus for theorising in evolutionary economics.

Knowledge, which forms the basis for competence, is either codifiable (and therefore readily transferred) or tacit (and therefore not readily transferable beyond the context in which it is embedded). Codifiable knowledge, in the form of scientific and other forms of scripted or formal knowledge, has an ubiquitous nature once access to its sources is mastered, but the entry barriers to new knowledge can be very high. Tacit knowledge, importantly, is specific to organisational and geographic locations and this increases its internal circulation but impedes its external accessibility. Learning processes, which absorb information and generate and diffuse knowledge (of both sorts), are collective activities that form part of the background and experience of each organisation. Their effectiveness depends in part on the quality of social interaction and lines of communication. These are enhanced by a shared social and cultural environment from which develops common routines, norms and standards, which depend upon, and foster, trust and the willingness to co-operate. The ability to form and maintain effective social relations is therefore a key competence.[2]

From this perspective, neo-classical theory can be criticised for treating knowledge as information that can be readily absorbed and for neglecting the fact that knowledge systems are required if information is to be used effectively (Lundvall and Borrás, 1998). The important point here is that it is not so much information that is scarce but rather the ability to locate and select the relevant information and to convert it into useable knowledge. Therefore, the analysis of knowledge diffusion and innovative capabilities necessitates the recognition of influences beyond such static notions as the availability of information and the given research, science and technology base of organisations at given points in time. It requires the recognition of the dynamic and evolving interplay between information, codifiable and tacit knowledge, and competence. The specificity of tacit knowledge and competencies means that externally derived information requires converting if it is to be readable within internal knowledge and learning systems. The absorption of radically new codifiable knowledge requires the development of new, or the modification of existing, tacit knowledge if competencies are to evolve effectively. The problems of the absorption of change may be eased if the creators and users of new knowledge share the relevant tacit knowledge so that effective interpretative mechanisms can develop. For any organisation, then, the successful generation, diffusion and utilisation of new ideas can be expected to involve interactions between internally and externally generated codifiable and tacit knowledge extending to the range of suppliers, customers and research institutions to which it relates. Nevertheless, although such close linkages are important for initiating and diffusing incremental change, they may also form obstacles to the spread of radical innovations requiring openness to the outside if the necessary fundamentally new competencies are to evolve.

Policy concern with these dynamic processes has become more and more important and urgent because intensifying competition and the pace and direction of technical change has increased the importance of learning and therefore the need for co-operation amongst firms and between business and knowledge institutions. And although accessibility to, and the speed of transmission of, codifiable knowledge has been radically increased by information technology, the crucial elements of knowledge remain tacit and rooted in specific organisations and locations so that the necessary learning processes to enable the effective incorporation of new knowledge put strict limits on how far codification can go. This has given further impetus for the sharing of tacit knowledge to facilitate technological and business networking and the forging of closer linkages between suppliers and customers; relationships which have extended internationally as multinational firms tap into localised specific knowledge and in the process become embedded in local networks. Therefore, perhaps paradoxically, the intensified competition which drives and is driven by technical change requires greater co-operation within and between firms, localities and knowledge centres. However, the more activities have become integrated and the more important has become locality, the greater the risks there are of lock-in. Moreover, generic change in technology has resulted in the emergence of new ways of organising production, which necessarily extend beyond traditional technological, industrial and institutional boundaries and require increased interaction between different scientific and technological specialisms. Such developments need openness to new forms of knowledge and the preparedness to realign networking.

This last point highlights the need to distinguish learning processes from the ability to adapt to radically changed technology (Amin and Wilkinson, 1999). This may require firms to shed old learning or established routines, practices, and outlooks, and the inability to do this may hinder the necessary adaptation to new technologies and forms of organisation. The ability to adapt is influenced by the strength of ties within and between organisations which may act as impediments to change. It is also influenced by prevailing managerial and operational cultures and conventions, which may or may not encourage the re-evaluation of past and existing practices in the light of new developments. Adaptation, in this sense, is centrally dependent upon the interplay between the learning propensities of firms and business networks, and the malleability of their organisational form. This raises an important dilemma for policy. For whilst closer integration, both within firms and within networks, to facilitate learning processes is an important objective, the risk is that organisational inertia will block access to radical change and lock productive systems into obsolete technological trajectories.

A second policy dilemma raised by the evolutionary and learning approaches to technical change is the possibility of a conflict between 'exploration' and 'exploitation', that is, between the development of radically new products and processes and the incremental incorporation of new technology into existing production systems (Lundvall and Borrás, 1998). In much of these discussions, however, the distinction is too sharply drawn because exploration and exploitation are essentially different stages in a continuous process. A more useful distinction, indicating this continuity, is that McArthur (1990) makes between 'newly emerging' and 'widely diffusing' technologies. The status of computing, electronic and new telecommunication systems has changed from 'newly emerging' to 'widely diffusing' over the past 25 years: Internet technology is currently making that transition, whilst bio-technology is still in the 'newly-emerging' stage.

Technical change can, therefore, be best understood as a process the beginning stage of which is the diffusion of radical new ideas from the science and technology base and the end stage of which is the wide diffusion of new products and processes. The incorporation of new knowledge into products and processes has been identified as a four-stage cycle the first of which is the building of shared values, norms and technical understanding so that often diverse knowledge can be shared (Lawson and Lorenz, 1999). To an important extent the second stage is crucial because it is at this stage that the cross-fertilisation between science, engineering, production, marketing and other specialisms is achieved. In this collective effort, it is essential that the contribution of each is sufficiently understandable to the others. The second stage is when individuals with diverse and complementary knowledge come together and collectively seek to explain their ideas about a new product or technology. This requires the members of the group to articulate early ideas about new developments by clarifying their notions and developing new concepts that are mutually comprehensible within the group. Modified in this way new knowledge becomes easier to combine with that of known technologies and methods in the process of building testable prototypes. At the fourth stage the new product or process goes into production and with this the knowledge underlying the new competencies, which was articulated in the initiation and development phases, becomes increasingly tacit and forms the basis for new knowledge creation by learning by doing and incremental technical change.

If the process of technical change is envisaged as a spectrum of activities with the production of new knowledge at one end and the production of new and improved products by new and improved processes at the other, the high-technology sector largely occupies the intermediate ground. The function of high-technology activity ranges from the conversion of new scientific and technical knowledge into a form which is accessible to industry, through the origination and proto-typing of new products and processes and the progressing of these to production for the market, to the provision of specialised scientific and technological services for both the high-technology and more traditional sectors. For any economy, the acquiring of the full value-added from technical development requires a balance between a high quality scientific and technology base, a high-technology sector, and more traditional industry, together with an institutional structure which supports technological diffusion and the necessary learning processes within and between these sectors.

9.3 Technological and Social Relations of Innovation

A major feature of the evolutionary/competence approach is the way it highlights the dynamics and essentially collaborative nature of technical development. In turn, this underlines the importance of both technical and social relations in evolving knowledge-learning-technology systems. Technical relations are the linkages by which different strands of new and existing knowledge are combined in the development of new products and processes. These are objective and functional relationships between the producers, developers and commercial exploiters of knowledge. They are inter-disciplinary both in the sense that they bridge the gap between pure and commercially applied scientific knowledge and in the sense that they bring together different types of both pure and applied knowledge. By contrast, the social relations of production are the subjective and personal associations between individuals and groups which form the social network within which the science and technology of production is advanced and processes of technological development and innovation are jointly undertaken. Social relations of production play a central role in determining the effectiveness of scientific and technical co-operation and hence the success of innovative systems.

The social relations of technology development have three functions in the management of innovation: direction, co-ordination and command. Direction involves such entrepreneurial decisions as choice over the final form innovation takes, the methods of introduction and marketing strategies; co-ordination synchronises the stages in the innovative process; and command exercises the control and imposes the sanctions necessary for co-ordination. Each of these functions will require a degree of formal authority and a network of more informal interpersonal relationships among participants in the innovative process, although this mix will vary between managerial systems. Together, they serve to secure effective co-operation and therefore work in the mutual interests of the partners by securing for them the best possible outcome in term of the largest aggregate return for their collective effort. From the evolutionary/competences theoretical perspective, technological and social relations interact in the determination of the nature and direction of technological trajectories, and both are subject to continuous pressure for change.

What receives little if any attention in these theories are the distributional dimensions of social relations. What is neglected is the reality that whilst the parties involved in the innovation process derive mutual benefits from co-operation, they necessarily compete over the proceeds because what one gets the others cannot have. This distinction is important because of the effect securing distributional advantage might have on the trajectory of technical development and because the failure to resolve distributional conflicts could result in the withdrawal of co-operation, a slowing down of technical change and a reduction of the competitive benefits of innovation. Such outcomes depend importantly on the bargaining strength of the parties involved. Problems of relative power are recognised by neo-classical theorists in their ambiguous attitude towards monopoly control of knowledge. On one hand, it overcomes market failure resulting from spill-over and provides the incentive to innovate whilst on the other hand, monopolisation is a source of market failure causing welfare loss by distorting distribution. No comparable attention has been paid to the effects of unequal bargaining power in the economic theories of evolution and competences.

Innovative systems, like the productive systems in which they are embedded,[3] are typified by mutual dependence and relative power and each relationship is by its nature both co-operative and rivalrous. Social relations have the dual and potentially conflicting roles of securing co-operation in the process of producing, and agreement over distribution of the outcome of, innovation. At the micro-level these social relations take the form of legally binding commercial, employment and other forms of contract, authority relationships within intra- and inter-organisational hierarchies based on power, and more informal linkages based on the recognition of mutual dependence. At the macro-level the social relations of technological development and innovation are structured by the political and legal systems. These regulate relationships by determining the form taken by commercial and other forms of contract, establishing the rules balancing the protection of intellectual property rights against the free flow of information, administering procedures for resolving disputes, regulating private organisations and institutions, and determining what constitutes breaches in rules and regulations and imposing sanctions. These rules, norms, procedures and sanctions are embedded in company law, competition policy, patent law, and labour, product and capital market legislation.

As Nooteboom (1999) emphasises, organisations consist of assets, competencies and positional advantage which to a great or lesser extent are organisationally specific. And while tangible assets are owned and contracted for, competencies and positional advantage are not. At the level of the firm, competencies refer to its technical and organisational capabilities, and positional advantage to its degree of monopoly, reputation and other advantages it might have in the markets and networks of firms and institutions in which it operates. The firm's specific competencies and positional advantage means that contracts are necessarily incomplete creating problems for monitoring performance and for the distribution of rent (Deakin and Slinger, 1997). Moreover, the trust and co-operation between contracting parties necessary for developing and maintaining competencies may importantly depend on how performance and distributional disputes are resolved (Wilkinson, 1997).

Positional advantages may also impact upon competences and the trajectory of technical development and hence dynamic efficiency. Batt and Darbishire (1998) argue that the form taken by the deregulation of telecommunications in the US and Britain dictated the path of technical development. It allowed telecommunications companies to deploy new technology to segment the product and labour markets to their distribution advantage and to extend this internationally by constructing global networks. The different regulatory framework in telecommunications in Germany, involving a wider range of stakeholders, resulted in markedly different patterns of technical change and quality of outcomes and a more even distribution of the benefits from technical change. In a similar way, the rivalry created by the use made by Microsoft of its market power and the outcome of the current anti-trust suit against Microsoft can be expected to have important consequences for innovation in information technology and for the structuring of the market for internet services. The legal regulation of technical change and political and legal constitution of markets therefore create frameworks within which competencies evolve, and have a central role in determining technological trajectories.

An innovation system can be therefore defined

> .... by both, historically specific technical and social interdependencies, where the notion of social interdependencies encompasses mutual interest (consensus) and relative power (contest). In this perspective, the endogenisation of social and technical change is not so much a matter of showing that societal institutions and the organisation of innovation arise as solutions to cognitive and economic constraints, but of analysing the political bargaining process underlying the emergence of particular social or technical arrangements or settlements in production (Blankenburg, 1998, p. 53).

The failure to make this distinction is apparent when comparisons are made between America, which is seen as taking the lead in radical new technology, and European countries and Japan, where innovation is seen as being more incremental. There is no doubt that the USA has taken the lead in the 'newly emerging' phases of most recent new technologies. However, much of the lead in incorporating these generic changes into productive systems has come from Japan and Europe rather than America. This lack of success by American business cannot be simply attributed to the fact that exploration beneficially crowded out exploitation. It can be more readily accounted for by failure of American industry to foster effective co-operative learning processes in the workplace, in supply chains, and with the science and technology base because of extreme individualism and antagonistic inter- and inter-firm relationships. It is also to be explained by the prevalence of Taylorist forms of work organisation with their emphasis on the de-skilling of the workforce, the concentration of conception and organisation of work in the hands of management, and the idea that skilled work can be readily replaced by technology. The corollary of this is a concentration of education and training at highest scientific, technological and managerial levels and a neglect of the middle range technical skill base which places limitations on organisational learning (Lazonick, 1997). This helps explain why, for example, America lost the world lead in the development and sale of machine tools, a primary route for the diffusion of electronics and computing into manufacturing, to Japan (Forrand, 1997). Thus, America's relative failure at the 'widely diffusing stage' can be explained by a failure to make radical innovations in work and industrial organisation and related policies to support its lead in radical new science and technology. This conclusion reinforces the argument that the lack of organisational flexibility will lessen the ability to meet the challenge of radical change. The locking-in of America business into a mode of rationality (Pratt, 1996) typified by Fordist work organisation, extreme individualism and antagonistic employment and business relationships inhibited learning capabilities (Lazonick, 1991) and served as a barrier to the effective exploitation of radically new knowledge from its scientific base. By contrast, in many European countries traditional organisation of the science and technology base proved more of a weak link. The central policy point, however, is that technical progress requires the capability and flexibility to both explore for and exploit new knowledge. This requires high quality basic research, the ability to identify and to develop economically viable products and processes, the development of a supportive political, organisational and social environment, and an industrial base flexible enough fully to exploit the new opportunities.[4]

The adoption of a broader theoretical perspective, which recognises the importance of both dynamic cognitive-technological and social/political developments, widens the policy agenda. Policy has to be adaptive to the challenge of new technological, organisational and market forms spawned by the pressure of scientific discovery and intensified competition. Such policy objectives include the adaptation and upgrading of technical and social support infrastructures to accommodate change by encouraging the development, and when necessary the reconfiguration, of co-operative networks, to enhance their innovative capacity. The effectiveness of these policies and their effect on the speed and direction of technical change and on competitive performance will also depend upon the devising of effective competition, industrial, regional and labour market policies. An important objective here should be the development of regulatory frameworks and codes of practice designed to restrict the exploitation of bargaining advantage and the negative consequence of this for the diffusion of new technology and the collaborative effort necessary to deploy it to its fullest advantage.

9.4 Policy Implications for High-Technology SME Innovation and Competitiveness

The work of the network lends empirical and theoretical support to the evolutionary/competence approach to innovation systems whilst warning of the need not to lose sight of their socio-political dimensions. A key feature of innovation systems demonstrated by the case studies is the increasingly central role of collaboration, interaction and networking between firms and other partners including universities/research organisations and institutions. Through these processes, information and knowledge is diffused and innovative capacity is enhanced. Notwithstanding the national and international orientation of much of this activity, for many small high-technology firms the geographical focus for knowledge access and distribution is the local area or region. It is here that proximity and a shared social and cultural environment provide the channels and means for the knowledge exchange, trust and co-operation necessary for collective learning to take place.

There is also a general awareness that restrictions on knowledge flows have become increasingly costly as the pressure to innovate has intensified. This has increased the importance of links between business and the science and technology base, links which are becoming increasingly complex with the growing need to combine different technologies and to acquire interdisciplinary capabilities. There is a growing need to shape the knowledge infrastructure and channels of technology transfer to meet changing requirements. This requires a reduction of supply-side constraints on the flow of information and knowledge, and an increase in the capabilities and motivation of SMEs to absorb and use effectively new science and technology. A complementary development is the need to foster closer relations between firms to encourage technology diffusion and collective learning. Although all this activity can and does occur without policy intervention, the case studies suggest that policy has an important part to play in promoting partnerships and creating channels for learning. The research also suggests that policies need a locality and industry specific orientation if they are to contribute to building the relationships of trust and confidence essential for effective networking (Malecki, 1997).

9.4.1 Diffusion of Knowledge from the Science and Technology Base

Whilst the imperative for universities and public research institutes to collaborate with business in R&D is now quite widely recognised, important barriers to such collaboration persist, particularly for small and medium sized enterprises. In this respect (Lundvall and Borrás, 1998), the distinction between functional and dysfunctional barriers to the transfer of knowledge to business from universities and other public research laboratories is particularly helpful. Distance between business and the funding of university research is functional in the scope it gives the latter for carrying out long-term widely disseminated basic research without the expectation of immediate commercial gain. But barriers may be dysfunctional if they hinder business access to useful new scientific and technical knowledge or restrict it to those firms with the in-built capacity to decode and incorporate into their knowledge base the results of leading edge scientific research. Supply-side obstacles may result from differences between academia and business in values, objectives and organisational structures which prove difficult to surmount. Demand-side obstruction stems from ignorance on the part of firms, especially smaller firms, of the availability of transferable knowledge or how best to use it. These problems of communication and knowledge transfer can be exacerbated by a multiplicity of intermediary institutions with overlapping responsibilities, and a failure effectively to identify the needs of target groups.

The network's research shows that commercialisation of the science base through small companies spinning out of universities and research institutes is a significant feature of all the successful European regional clusters studied. The case studies show that the nature and extent of the spin-offs varies from country to country and reflect, among other things, the scientific and technological strengths of the local research community, the institutional and legal arrangements with respect to intellectual property, and the socio-economic context supporting the spin-off process. Spin-offs were almost invariably found to locate in the same area as the parent organisation and thus contribute to the development of the local high-technology cluster. There is also evidence of a wider process of knowledge diffusion and learning as academic spin-offs collaborate with other firms in the local cluster, and as they themselves spawn new high-technology start-ups.

The case studies suggest that the potential for this spread effect is enhanced by a sharing of specialisations, technical capabilities and values between knowledge institutions and local industry. Such close links are to be found in Göteborg, Helsinki, Munich and Genoble, all of which have high quality and research-active technological universities. It should also be noted that these are industrial regions which have been particularly successful in economic regeneration. This can be explained by the high quality and applied orientation of the research which together with the links with local industry have opened up the areas to new technology and have helped prevent industrial and organisational lock-in. This also helps account for the effectiveness of these areas in generating spin-offs, mixing technology and evolving synergies with local industries which allow more of the value-added to be trapped in the region. Comparisons can be usefully made with the Cambridge case where research is less applied and where there is less of an established industrial base. The rapid growth in the number of high-technology firms in the Cambridge region suggests the emergence of an industrial hinterland to the University, although its research boundary with local firms is closer to the 'pure' end of the research spectrum than it is in areas with top-flight technological universities. Moreover, the customer base of Cambridge is more external to the region than in areas with more diversified industrial structures.

As well as technological and sectoral affinity, close geographical proximity between industries and academic research centres increases the prospects for interaction, spin-offs and development of collective learning processes. In Sophia-Antipolis, for example, local interactions were sparse before the research departments of the University of Nice relocated to the park. This made possible increased interaction resulting in a progressive increase in the number of small-firm spin-offs from the university. Within a few years this had a cumulative effect as more and more SMEs were created and as the local qualified labour market expanded, enabling increasing cultural, personal and technical networking between graduates in the university and in industry.

In addition to the important role SMEs are playing in diffusing new technology, there is also growing evidence of a sea-change in the strategies of large multinational firms towards universities and public sector research institutes. The growing need for continuous technological change is requiring them to tap into localised and specialised R&D concentrations (in the form of knowledge centres and clusters of high-technology firms). This has involved the relocation of research laboratories to high-technology regions and encouragement to subsidiaries to embed themselves in localised technology-based clusters. The greater involvement of large firms can be seen as a means of introducing new impetus into localised clusters helping avoid the risk of *lock-in* and accelerating development by providing a source of new technical and organisational knowledge. This is diffused by spin-offs, by labour mobility within an extended highly-skilled labour market, and by the incorporation of large firms in local networks by the development of close links with local small and medium sized firms. There is considerable evidence of these positive effects in regions such as Munich, Göteborg and Utrecht where close links between large and small firms (including those between spin-offs and their parents) benefit both and contribute substantially to technological diffusion and learning processes within the regions.

High-technology consultancies can also act as an important bridge between the science and technology base and industry, and Cambridge is notable as the location of a large number of high-technology business service firms specialising in the production of technological knowledge and competence (Lawson, 2000). Like technological universities and, in some areas, large firms, technology consultancy firms serve as links between 'pure' science and innovation. They intermediate between firms and the scientific community by providing 'quasi-generic' knowledge extracted by means of repeated interactions with their customers and the scientific community. They are also themselves major sources of new high-technology spin-offs. High-technology consultancies therefore provide a conduit for incorporating scientific knowledge into tacit knowledge, learning processes and firm competencies (Antonelli, 1999). The growth in international high-technology consultancy contributes to the creation and distribution of both tacit and codified knowledge at the world level and in doing so consultancies help both to overcome the insularity of local networks and widely to diffuse local knowledge. Their services are much more carefully tailored to the requirements of their clients than much technology transfer directly from universities. However, this means that the information is much more private and less widely available.

Finally, there is the important role played by government expenditure in triggering new technologies and diffusing them widely. Space and military expenditure played a key role in the USA in the promotion, development and diffusion of electronic, computer and information technology. In Munich the concentration of military expenditure was also pivotal in encouraging high-technology development whilst in Sophia-Antipolis the relocation of prestigious government research centres and parts of the space programme was an important catalyst to the early stage of growth. Important non-military areas where policy intervention is necessary and the prospects for technical change real are in energy conservation and environment. There are also strong arguments for targeted procurement designed to facilitate building systems for delivering technology to industry and particularly small firms, and to increase small firms' ability to absorb knowledge and to innovate. This would have the added advantage of developing and providing socially, ecologically or technologically desirable innovative products for which there are currently few market incentives.

An important element of public policy aimed at raising the innovative capacity of business, therefore, is to facilitate the transfer of knowledge between academia and industry and encourage its exploitation and commercialisation. This may be achieved by a variety of routes including the direct transfer of information, the spinning-off of small businesses, inter-mediation by high-technology consultants, graduate recruitment, and the movement of scientists and technologists between universities and industry, including student placements. One of the most successful examples of the last of these is Germany's dual-training system, and the widespread undertaking by graduates and undergraduates of projects involving the industrial application of science and technology. Small firms have a particular problem in accessing, decoding and making use of new scientific knowledge. Potentially, high-technology consultancies also have a valuable part to play in overcoming the difficulties faced by small businesses. Currently, however, such consultants are much more frequently used by large companies in advanced areas, rather than by small firms and firms in less developed regions. There is a clear case for subsidising the use of technical consultancies by SMEs. But success in this respect requires quality control to ensure that the advice on offer meets the small firm's requirements, as well as careful preparation to ensure that the services offered by technical consultants can be readily absorbed by the recipients. A policy priority is co-ordination amongst possibly competing institutions engaged in the framing and implementation of policy, and transparency in this process to increase accessibility, especially for SMEs.

9.4.2 Proximity, Networking and Collective Learning

Involvement in collective learning processes can be seen as a means of overcoming both supply and demand-side obstacles to converting new scientific information into innovation capabilities. As noted in chapter 1, regional collective learning can be defined as involving 'the creation and further development of a base of common or shared knowledge among individuals making up a productive system which allows them to co-ordinate their actions in the resolution of the technological and organisational problems they confront' (Lorenz, 1996). The successful regional clusters studied by the European research network have over time developed a collective learning capacity for creating, diffusing and elaborating new technological and organisational knowledge within and between the cluster's constituent firms.

These localised collective activities may be deliberate actions or 'joint-products' of other activities. Examples of the former are research collaboration between local SMEs, between an SME and a local university, and between an SME and a local large firm: examples of the latter include the movement of 'embodied expertise' as researchers, managers and skilled workers move within the regional labour market and as they spin-off new small firms. The links firms establish with other firms and with research institutions are both formal and informal. The latter play a very important role in technology transfer and collective learning. These are developed and maintained by training links between universities and firms (the dual training system in Munich being a good example) and by continuing relationships between spin-offs and their parent research institutions and firms found, for example, in Göteborg, Cambridge, Linköping, Grenoble, Munich and Sophia-Antipolis. Linkages of varying degrees of formality, by which information and knowledge is diffused, have been developed through and supported by a variety of social and institutional infrastructures, including trade, industry and professional associations, business clubs, local partnerships and development agencies. In addition, in several clusters new collective enterprise initiatives have emerged which are engaging in marketing, publicity, networking and business support, and which provide further opportunities for the development of collaboration between firms and universities. These are usually private-public coalitions, as in Cambridge, Sophia-Antipolis and Grenoble, and can involve business and professional associations, local government, public business support and training agencies, universities and public research institutes.

Admittedly, geographical clustering and its numerous benefits to firms in the cluster are inadequate on their own to ensure continuing successful innovation and firm growth. Firms also need to be involved in wider national and global networks because regional collective learning and wider networking are complementary and mutually reinforcing, rather than alternative, processes (Camagni, 1991, p.139). Both are necessary for high-technology SMEs to sustain their innovative activity and competitive advantage. Most of the leading high-technology SMEs in the Oxford and Cambridge milieux are engaged in a variety of international collaborative initiatives with firms/universities in other networks, activities which are associated with above-average performance. It follows that business support agencies concerned with technology-intensive SMEs should be sensitive not only to the importance of local interactions, relationships and networking, but also to the establishment of such links with firms and organisations in other local or regional networks.

Perhaps the key conclusion from the case studies, however, and one with strong policy implications, is that geographical and institutional proximity is important for the development of collective learning, and that location in a 'high-learning' cluster strengthens the innovative capacity of both small and large high-technology firms. In particular, this provides support for a spatially focused network approach to innovation policy. Such policy should seek to encourage local SME networking and remove constraints on collective learning and resultant enhanced innovativeness. Collective learning constraints may relate to limitations on spin-offs from companies and universities, on inter-firm networking and collaboration, and on labour mobility. In a number of the clusters studied, an important role is played in lifting these constraints by local and regionally based institutions which support the creation and diffusion of knowledge. There is therefore a potential role for policy in creating or supporting such institutions where they are poorly developed or absent.

Nevertheless, networking poses complex policy problems. Policy may contribute to the *collective capital* (Dei Ottati, 1994) for fostering trust and collective learning by establishing a framework of rules and norms regulating the relationships between firms and their links with research institutions, by sponsoring organisations and institutions which support and service SMEs, and by bringing groups together (for example, as part of procurement policy). But even within this supporting framework, the building of *personal capital* (Dei Ottati, 1994) for high quality relations may require considerable time and experience before the necessary trust and reputation can be created (Lorenz, 1988). Consistency in policy will be important in creating this long-term perspective although, because of the risks of *lock-in,* policy makers should be alert to the need to promote new networking opportunities and the need for external (especially, international) linkages.

9.4.3 Business Support Policies for High-Technology SMEs

Important policy implications arising from the network's research with regard to high-technology SME support policies are firstly, the need for local support agencies and institutions which actively interact in the process of collective learning, particularly in identifying and resolving problems confronted by businesses within a particular local technology-intensive production system; and secondly, the need to understand how business support policies can be tailored to meet the specific requirements of such firms, including their exceptional needs for highly-qualified R&D, technical and managerial staff.

Although there would seem to be a general case for government policy encouraging and supporting high-technology SMEs in order to facilitate the exploitation and diffusion of knowledge, the case studies indicate that policy needs to based on a clear understanding of the local production system. There are striking contrasts, for example, between the development of spin-off activity in the different regions and this confirms that there are no simple patterns of behaviour or policy responses but rather that policy must reflect local conditions, constraints and practices if it is to meet the specific needs of technology-based spin-offs in each area.

That said, however, the network's research also supports other recent findings (for example, Moore and Hughes, 1998) that in certain respects, small high-technology firms tend to face more severe barriers than more conventional SMEs, both at start-up and during subsequent growth and development phases. Thus small high-technology firms are more likely than other small firms to face constraints in marketing and sales skills, accessing finance, and finding suitable premises. Their much greater propensity – and need – to collaborate with other technology-based firms and organisations also suggests another area where business support agencies may be helpful in identifying and enabling inter-firm partnership opportunities. A particularly interesting finding reported to the European network by Autio (1997: see also chapter 4) is that in the Helsinki case, local high-technology SMEs clearly prefer provision of specialised, customised and hands-on technology support services rather than more general 'infrastructural' technology services provided by 'centralised middle-men'. The former also require 'longterm interaction between the service provider and the recipient of the service' (Autio, 1997, p. 64). Autio suggests that this indicates a need for the provision of local 'dedicated technology brokers' to facilitate interaction on a longterm basis.

Education and training policies targeted at their specific needs are also crucial elements of business support packages for firms in high-technology clusters. The widening of the range of specialised qualifications and technological expertise available in local labour markets is an important endowment to support new start-ups and established firms as well as to attract incoming firms. A second important need is for training in new organisational methods and approaches. Demands being made for closer co-operation, improved external and internal communications, more effective control of quality, problem solving within teams and extensive use of information technology, require higher levels and wider ranges of social, scientific and technical skills. But it is not only the content of educational and training programmes that are important but also how they are organised. High level technical education can be an effective route for interchange of information between universities and firms and for building technology transmission systems. Examples from the case studies include the dual training system in Munich, the use of graduates for projects by small firms in Helsinki, and the use of managers from high-technology firms for teaching in Grenoble's universities. Each of these schemes underlines the importance of the organisation of training and of tailoring training to the needs of local industry. This is important not only for providing appropriately skilled workers but also in building the capability for translating new knowledge into usable forms, for building capacity for collective learning and creating credible delivery systems for knowledge diffusion. Moreover, policies which emphasise the importance of co-operation between industry, research institutions and the trainers of highly skilled workers are essential for building training capabilities ahead of changing demand.

9.4.4 Regional Development Policies

Raising regional innovative capability, strengthening the knowledge base of the region and improving the learning capacity of the region are now familiar themes in the development of regional and local development policies. In this important context, the evidence from the case studies relating to collective learning processes, networking and inter-firm links and the role of support given to these by regional and local institutions has important implications for regional and local policies. This lesson has already been learned in a number of EU countries which have seen a shift away from top-down regional/local policies, with their emphasis on investment incentives and demand re-distribution, towards policies in which regional organisations have greater autonomy and are an integral part of the collective learning process. In addition, there is now widespread interest at both national and regional levels in strategies supporting the development of regional 'clusters' of firms engaged in formal and informal vertical and horizontal interactions (see, for example, Secretary of State for Trade and Industry, 1998, pp. 45–6). This changing orientation of policy is strongly supported by the evidence from the network on the importance of regional clustering of high-technology SMEs in technology transfer and collective learning.

The case studies suggest that policies formulated and implemented at the local and regional level are likely to be most effective in addressing the unmet business development needs of university spin-offs and other local high-technology SMEs. It is also at the regional and local level that policy initiatives supporting collaboration between local high-technology SMEs and local knowledge institutions are best formulated and implemented. The experience of Grenoble, for example, provides evidence of the benefits of intra-regional linkages between SMEs and research institutions, as well as of the supportive role which can be played by regional and local government. Regional and local planning and economic development policies can help raise innovative capability by ensuring the availability of suitable sites and business premises close to universities/research institutes integrated with or in close proximity to business support services with experience of the business development needs of small high-technology start-ups. The experience of Cambridge, Oxford and Munich in the evolution of their local planning policies, and the development of science parks and innovation centres in a number of the other regional clusters studied by the network, demonstrate the importance of sympathetic and sensitive local planning and business support policies in creating conditions for successful spin-offs. The need to encourage local university participation is emphasised by the Helsinki experience where policy integrates entrepreneurship training, admission to small firm incubator schemes and the provision of technical and business services.

The case studies also demonstrate that sources of competitive advantage are mutually reinforcing and as a consequence high-technology clusters typically enjoy relatively rapid economic growth. This implies that regional strategies for such clusters must be designed with a view to accommodating growth and avoiding congestion of public services and infrastructure. On the other hand, the strong tendency for effective commercialisation of the science base and the development of high-technology clusters to be concentrated in already prosperous European regions carries with it the danger of persistent and possibly increasing disparities between regions with and without rapid rates of high-technology development. An important question is whether policies concerned with fostering the development of high-technology clusters, and of strengthening the science base of and firm links within such clusters, may in time lead to *trickle across* to less well-endowed regions. The case studies do reveal that relatively few regions in each country have developed high-technology clusters and strong local academic/industry links, and suggest that it is unrealistic to expect that more than a few regions will be able successfully to pursue this regional economic evolutionary trajectory. Admittedly, the experience of Sophia-Antipolis does show that policy-led initiatives can sometimes be successful in initiating and growing high-technology clusters in a peripheral region without a high-technology/R&D tradition. However, the Sophia-Antipolis case also demonstrates the very substantial nature of such initiatives, the fundamental rather than marginal changes required, the need for long term commitment and resources, and the length of time needed before objectives are fully met. In addition, few peripheral regions of Europe offer the residential and environmental attractiveness to highly qualified workers and entrepreneurs, as well as accessibility by air, of this particular location.

Where high-technology cluster promotion is not feasible, a more relevant policy message for technologically disadvantaged regions is probably to recognise the benefits of strengthening links between local universities and research institutes and local industries. Such a development has been recently mooted for the West Midlands region of the UK where 'traditional manufacturing industries are at the heart of the scheme' (*Financial Times*, 20 October 1999). The lesson of the case studies reported in this book is that the synergy between new technology and traditional industry is most effectively forged where there are top flight technology universities which have built up close research and educational links with local industry. In addition, more traditional regional policies aimed at attracting inward investment as a source of regional development are more likely to be successful if they are able to target companies with specific R&D needs which match the research/technological strengths of both local universities/research institutes and local SMEs.

It is apparent that top-down policies to address regional disparities are unlikely to be sufficient and that without regional and local mechanisms to ensure take-up, national programmes will benefit areas with established innovation systems and technology transfer mechanisms. Such policies cannot create the formal and informal interactions identified in the case studies which encourage the commercialisation of the local science base. The importance of proximity for collective learning processes and for innovation networks in strengthening innovative capacity underlines the need for a regional dimension to innovation policy. It is at this level that the knowledge of the need for effective technology policy and how it might be implemented is most likely to be found. Consequently, it is at the regional level that the funds available can be most effectively deployed, especially for SMEs. On the other hand, 'regional innovation policies cannot replace national policies for techno-industrial innovation' (Hilpert, 1991). And moreover, funding sources for policy are usually much wider than regional, and the availability of funds is much greater from outside the region than inside. Concerted action involving local, regional, national, and European levels is therefore necessary, although such broad participation is unusual. It has been most effectively implemented in Grenoble and least effectively in the Oxford and Cambridge clusters because of a lack of local autonomy. In other regions, promising attempts at co-ordination of policy have been made (for example, Helsinki, Milan, parts of the Randstad). The central purpose of these integrated approaches is to tap into national and international initiatives and funding, and to build local and regional competences which are open to outside influences.

9.4.5 Socio-Political Dimensions of Policy: Large Firms, Competition and Intellectual Property

So far the discussion has been confined to the scope for policy intervention to improve the cognitive/technical performance of regional innovation systems. This section is more directly concerned with socio/political relations within such systems and the implications of these for power relations, distribution and performance.

From the perspective of their relative bargaining power, the growing involvement of large firms with universities can be seen as a way by which they can increase their control over research and development, exclude other firms from such involvement, secure privileged access to pools of scarce highly skilled labour, and increase their dominance over small and medium sized businesses. These considerations raise difficult policy issues for whilst there may be benefits from encouraging the insertion of large firms and involving them in local clusters these need to be balanced against the risk of exploitation, monopolisation and the potential risks to knowledge diffusion of large firms attaching strings to the funding of basic research.

Dominance by large firms may also limit the opportunities for exploiting technological innovations open to smaller high-technology businesses and leave them with few options other than selling out to a large firm and by so doing capitalising the future potential of their businesses. The internalisation of innovative capability within large companies no doubt has its benefits in terms of the effective funding of research and development and the strengthening of the capabilities of the acquiring firms (see chapter 6). On the other hand, it limits the range of possibilities for innovation and learning to those bounded by the firm's technological trajectory. Acquisition of small firms is a way by which large firms remove potential rivals and reinforce their monopoly position by securing patented and unpatented knowledge and research capability. This raises difficult policy problems of trading the benefits of innovative activities of large firms against the cost in both distributional and technical progress terms of the exploitation of their market power. This has implications for competition, stock market and corporate governance policy. It also suggests a need for policy initiatives which encourage small firms to consider and adopt alternative growth strategies and sources of research collaboration and funding so as to provide credible alternatives to being acquired by large firms.

A closely related question is the role of competition policy in restricting or encouraging co-operation between firms. Until recently the importance of such co-operation was either neglected in the formulation of competition policy or regarded as a potential threat to competition and hence economic efficiency. What has changed this is the now undoubted and growing importance of inter-firm collaboration in technological development (see chapter 2, section 2.7). From an orthodox perspective, inter-firm collaboration is seen as a way round the problem of under-investment because of spill-over effects. Inter-firm alliances also offer the possibility of synergies in research and development, risk sharing and avoiding the possible rigidity and obstacles to the flow of knowledge from vertical integration. From a more dynamic perspective, the blocking of inter-firm alliances by competition policy encourages exclusively in-house innovation and a merging of would-be co-operators. This risks increasing concentration and impeding of innovative activity by restricting the open diffusion of new technical knowledge and by limiting collective learning.

Patenting and other means of protecting intellectual property rights also have conflicting effects on the pace of innovation. On one hand, they increase the certainty of appropriating the returns to invention and innovation. But on the other hand, they restrict the diffusion of radically new knowledge and innovative technology (Antonelli, 1999). These latter effects are particularly important at the level of the diffusion of basic research findings, limitation of access to which reduces the probability of innovation and the creation of the inter-firm networks necessary to bring these to fruition. This is the core of the argument for the adequate public funding of basic research, together with the argument against private funding which similarly risks restricting access. The network's case studies suggest that technological dynamism is linked to free access to the findings of basic research. In particular, the ability of successful university or institute researchers to exploit commercially the results of their research provides the incentive for spinning-out firms. Any move by research institutions to impose property right restrictions on such business creation activity risks slowing the diffusion of knowledge and slowing technical advance. Again there is a trade-off between the adequate funding of research and the effectiveness of exploiting the output from basic research.

In more general terms patenting and other restrictions on the use of radically new ideas are difficult to enforce. This is particularly so for small firms because of the cost of defending their rights at law. This puts them at a disadvantage compared with large firms who can more effectively defend their patents and use them to restrict entry.

9.5 Concluding Reflections

An important feature of most of the high-technology innovation systems studied by the network is the diffusion of information from the local science and technology base and its conversion into new products and processes with commercial value. Scientific and technical knowledge is directly transmitted by the dissemination of research findings or more indirectly diffused as embodied knowledge by industry's recruitment of graduates and post-graduates, intra- and inter- organisational labour mobility, the spin-out of new firms, and the work of high-technology consultants. Collective learning plays a central role in the combining of new and existing knowledge and in the creation of new, or significantly changed, products, processes and organisational approaches and structures. Collective learning involves collaboration between different scientific and technological disciplines, the producers and the commercial user of new knowledge, and between customers and suppliers. This co-operation is secured by close networking and its effectiveness requires the building up of shared experience and knowledge which enables the transfer of new ideas and their incorporation into the competences of innovators. The capability of achieving these objectives depends on the quality of relationships within the innovation chain, a mutual understanding of the technical, organisational and other needs, and a technical and attitudinal openness to new knowledge. The generation of scientific knowledge, its diffusion and collective learning, all have important implications for policy.

As pointed out above (section 9.4.1), the links between knowledge centres and industry can be either dysfunctional or functional. Reliance on business for funding risks biasing research and preventing the widest possible diffusion of its findings. There are therefore strong policy reasons for public provision of adequate core funding for leading-edge basic research and the widest possible dissemination of its results. On the other hand, close links between universities and industry are of central importance both for enhancing innovative capabilities and furthering research. These mutually beneficial, symbiotic relations are enhanced by geographical and techno-scientific proximity and the importance of this two-way relationship is underlined by the particular success of regions with strong science/technology research institutions, the specialisms of which overlap with those of clusters of local firms. In several such regions, this dual presence is proving vital for economic regeneration. Universities and public research institutes, with their more general interests, linkages into external scientific research networks and their closeness to local industry, are playing a key role in diffusing generic new technology to local industry and in implanting new technical specialisms. Close links between industry and universities are also important for providing local firms with appropriately educated and trained scientists and technologists and for forward planning to anticipate future requirements. These education and technology diffusion functions are also well served by designing undergraduate and graduate programmes which directly involve local firms. Well-established links are also invaluable in attracting outside funding for joint university/industry projects and success in such initiatives has been important for building effective local networks.

The importance of collective learning for building innovation capabilities means that there is a growing need for policies to evolve away from traditional notions of linear technology transfer towards policies of encouraging more interaction between research institutions and industry and a spatially focused network approach to innovation policy. Linked to this is the need for measures to encourage supportive local and regional organisations and institutions, and their continuous upgrading, to provide the broad and changing range of services needed, especially for small and medium sized high-technology companies. The greatest policy success is found where support services are tailored to firms' requirements and, again, where special efforts have been made to create a close working relationship between the teaching and research of local universities and the technical and other needs of local small firms.

The importance of proximity for collective learning processes and for innovation networks in strengthening innovative capacity underlines the need for a regional dimension to innovation policy. But concerted action involving local, regional, national, and European levels is also necessary, although such broad participation is unusual. The central purpose of these integrated approaches is to tap into national and international initiatives and funding and to build local and regional competences which are open to outside influences.

The national and European dimension is also important for addressing regional disparities in technical developments. High-technology development is highly concentrated in particular localities and the organic nature of the developments, and the long lead time and levels of resource required to trigger the growth of high-technology clusters, warn against any early, if any, success in reducing regional disparities by policy attempts to transplant high-technology activity. This means that any spread effect is more likely to come from more closely linking high-technology regions with those with more traditional industrial structures. The most effective route for this linkage could well be through local universities and research institutes (particularly those with strong technical orientations) and by building up links between these and local industries.

The potential for any productive system to innovate will importantly depend on the novelty and quality of its research output, the more widely it is diffused and the effectiveness of its collective learning processes which depend heavily on co-operative relations. But however co-operative the innovation process might need to be, the commercial exploitation of innovations is a highly competitive process in which the rewards, and hence the incentive to innovate, depend on the extent of the lead firms can secure over their rivals. The securing of this advantage may depend on the effectiveness of the innovation system to which entrepreneurs have access (systemic advantage) and/or their ability to exclude others from the knowledge required for innovation (monopoly advantage). Systemic advantage gives insider advantages over outsiders by means of the technical lead of the productive system and club membership. However, the ability to take up these advantages will depend on the entrepreneurial qualities of individuals. Monopoly over new knowledge gives firms advantages over insiders and outsiders but limits the diffusion of knowledge and may, by creating inequality in bargaining power and distribution, reduce the potential for collective learning and therefore reduce systemic advantage.

The ambivalent role of both collusion and individualism in both creating the capabilities and incentives to innovate and the opportunities to use that advantage for exploitative purposes lies at the heart of the dilemma for competition and intellectual property rights policies. Competition law has traditionally opposed collaboration between firms on the grounds that it is to the distributional disadvantage of outsiders and therefore results in the loss of economic welfare. The judgement has been less unfavourable to large size on the grounds that market success, and hence the growth of firms, results from the ability to innovative; and that the mass resources of large firms and the protection their dominance provides against technology spill-over provides both the means and incentive to innovate. Therefore, although large size creates distributional inequality, this is a price worth paying for a faster rate of technical change.

The growing recognition of the importance of close inter-firm linkages for collective learning and innovation has led to a questioning of the traditional attitude of competition policy towards inter-firm collaboration.

The benefits of industrial concentration for technical change are also being questioned on the grounds that the monopoly control of innovation capabilities, business funding of university research, and the dominance of small businesses by large firms can serve to buttress the monopoly power of big business, bias research, prevent the free flow of knowledge and restrict collective learning. This suggests the need for a redirection of competition policy which recognises the benefits of inter-firm collaboration whilst guarding more effectively against the negative effects on innovation capability of industrial concentration.

Similarly, protecting intellectual property rights in scientific knowledge by patenting and other means by individuals, firms, and universities has opposing effects by providing incentives to invent and innovate whilst at the same time restricting the diffusion of new technology. However, patenting and other means of securing property rights in new knowledge (by, for example, restrictive clauses in contracts preventing employees developing discoveries they have made for their own benefit) are notably difficult to enforce. This is particularly so for individuals and small firms because of the cost of defending their rights at law. This puts them at a disadvantage with large firms who can more effectively establish and defend their property rights and patents. In a period of rapid globalisation of research effort, these considerations raise important issues for national and international policy on patenting and intellectual property rights.

More generally, there is clear need for a co-ordinated approach to competition, intellectual property rights, industrial, regional, educational and innovation policy. Policy makers also need to bear in mind that the political and legal regulation of innovation and the factors affecting it form part of the framework within which competencies evolve, and have a central role in determining the trajectory of technical change. For both reasons, the long term objectives of policy need to be carefully thought out and clearly stated.

Notes

1   There is now a vast literature on these theories. For a valuable guide, and one which links these and the neo-classical theory literature directly to technology transfer, see Blankenburg (1998).
2   For extended discussions of these issues see the 1999 special issue of the *Cambridge Journal of Economics*, vol. 23, no. 2, on Learning, Proximity and Industrial Performance.

3   For the productive system approach see Wilkinson (1983), Biracree, Konzelmann and Wilkinson (1997), and Burchell and Wilkinson (1997).
4   A glance at the patterns of household consumption in Britain would suggest that high-technology applications have greater potential for product and process improvement in traditional areas of consumption than in relation to new high-technology products. In 1997, food, drink, tobacco, clothing, footwear, utilities, household furnishing and maintenance, and housing costs made up 50 per cent of consumer expenditure. Transport, recreation and culture and hospitality (restaurants and hotels) added a further 33 per cent. The areas of expenditure most closely associated with high-technology products, communications, audio-visual, photographic and information processing equipment, absorbed 3.5 per cent of household expenditure.

References

Amin, A. and Wilkinson, F. (1999), 'Learning, Proximity and Industrial Performance: an Introduction', *Cambridge Journal of Economics*, vol. 23, pp. 121–5.
Antonelli, C. (1999), 'The Evolution of the Industrial Organisation of the Production of Knowledge', *Cambridge Journal of Economics*, vol. 23, pp. 243–60.
Autio, E. (1997), 'University Links and Technology-Based SMEs in the Helsinki Region', in D. Keeble and C. Lawson (eds), *University Research Links and Spin-Offs in the Evolution of Regional Clusters of High-Technology SMEs in Europe*, ESRC Centre for Business Research, University of Cambridge, Cambridge, pp. 51–82.
Batt, R. and Darbishire, O. (1997), 'Institutional Determinants of Deregulation and Restructuring in Telecommunications: Britain, Germany and the United States Compared', *International Contributions to Labour Studies*, vol. 7, pp. 59–80.
Biracree, A., Konzelmann, S. and Wilkinson, F. (1997), 'Productive Systems, Competitive Pressures, Strategic Choices and Work Organisation: An Introduction', *International Contributions to Labour Studies*, vol. 7, pp. 3–17.
Blankenburg, S. (1998), 'University-Industry Relations, Innovation and Power: A Theoretical Framework for the Study of Technology Transfer from the Science Base', *ESRC Centre for Business Research, University of Cambridge Working Paper*, 102.
Burchell, B. and Wilkinson, F. (1997), 'Trust, Business Relationships and the Contractual Environment', *Cambridge Journal of Economics*, vol. 21, pp. 217–37.
Camagni, R. (1991), 'Local 'Milieu', Uncertainty and Innovation Networks: Towards a New Dynamic Theory of Economic Space', in R. Camagni (ed.), *Innovation Networks: Spatial Perspectives*, Belhaven Press, London, pp. 121–42.
Deakin, S. and Slinger G. (1997), 'Hostile Takeovers, Corporate Law and the Theory of the Firm', *ESRC Centre for Business Research, University of Cambridge Working Paper* 56.
Dei Ottati, G. (1994), 'Trust, Interlinking Transactions and Credit in Industrial Districts', *Cambridge Journal of Economics*, vol. 18, pp. 529–46.
Forrand, R. (1997), 'The Cutting Edge Dulled: The Post-Second World War Decline of the United States Machine Tool Industry', *International Contributions to Labour Studies*, vol. 7, pp. 37–58.
Hilpert, U. (ed.) (1991), *Regional Innovation and Decentralisation: High Tech Industry and Government Policy*, Routledge, London.

Lawson, C. (2000), 'Technical Consultancies and Regional Competences: the Case of Cambridge', ESRC Centre for Business Research, University of Cambridge, mimeo.

Lawson, C. and Lorenz, E. (1999), 'Collective Learning, Tacit Knowledge and Regional Innovative Capacity', *Regional Studies*, vol. 33, pp. 305–17.

Lazonick, W. (1991), *Business Organisation and the Myth of the Market*, Cambridge University Press, Cambridge.

Lazonick, W. (1997), 'Organisation Learning and International Competition: The Skill Base Hypothesis', *The Jerome Levy Economics Institute, Working Paper* No. 201, Annadale-on-Hudson, New York State.

Lorenz E. (1988), 'Neither Friends Nor Strangers: Informal Relations of Subcontracting in French Industry', in D. Gambetta (ed.), *Trust: Making and Breaking Co-operative Relations*, Blackwells, Oxford, pp. 194–210.

Lorenz, E. (1996), 'Collective Learning Processes and the Regional Labour Market', unpublished research note, European Network on Networks, Collective Learning and RTD in Regionally-Clustered High-Technology SMEs.

Lundvall, B-A. and Borrás, S. (1998), *The Globalising Learning Economy: Implications for Innovation Policy*, report to European Commission Directorate General XII, Office for Official Publications of the European Communities, Luxembourg.

Malecki, E.J. (1997), *Technology and Economic Development: The Dynamics of Local, Regional and National Competitiveness*, Longman, Harlow.

McArthur, R. (1990), 'Replacing the Concept of High Technology: Towards a Diffusion-Based Approach', *Environment and Planning A*, vol. 22, pp. 811–28.

Moore, B. and Hughes, A. (1998), 'High-Tech Firms: Market Position, Innovative Performance and Access to Finance', in A. Cosh and A. Hughes (eds), *Enterprise Britain: Growth, Innovation and Public Policy in the Small and Medium Sized Enterprise Sector 1994–97*, ESRC Centre for Business Research, University of Cambridge, Cambridge, pp. 86–98.

Nooteboom, B. (1999), 'Innovation, Learning and Industrial Organisation', *Cambridge Journal of Economics*, vol. 23, pp. 127–50.

Pratt, A. (1996), 'The Emerging Shape and Form of Innovation Networks and Institutions', in J. Simmie (ed.), *Innovations, Networks and Learning Regions*, Jessica Kingsley, London, pp. 124–36.

Secretary of State for Trade and Industry (1998), *Our Competitive Future: Building the Knowledge Driven Economy*, Cm 4176, HMSO, London.

Wilkinson, F. (1983), 'Productive Systems', *Cambridge Journal of Economics*, vol. 7, pp. 413–29.

Wilkinson, F. (1997), 'Cooperation, the Organisation of Work and Competitiveness', *ESRC Centre for Business Research, University of Cambridge Working Paper* 85.

Index

References from Notes are indicated by 'n' after page reference.

acquisitions 5, 42, 101, 156–78 (especially 174–6), 252
agglomeration economies 28
atomic energy 107–9, 119

Barcelona region 31, 37, 75–6, 94, 168, 173, 176, 204, 207–8, 219, 226n
Berkshire 8
biotechnology 6, 41, 44–5, 81, 87, 99, 107, 111, 165, 169, 175, 209, 234
business failures; see failures
business services, support, advice 23, 48, 50, 169, 215, 224–6, 247–51, 255

Cambridge Phenomenon 34, 99–100
Cambridge region 12–14, 31, 34–5, 39, 41–3, 49–51, 71, 76–8, 123, 126, 131, 133, 164, 168–9, 172, 176–7, 204–5, 208–11, 215–19, 222–5, 226n, 242–3
Cambridge university 13, 76–8, 96, 99–100, 103–5, 169, 205, 215, 222, 226n, 242
capital productivity 137–8, 154
Catalonia; see Barcelona region
Chalmers University of Technology 34, 43, 64, 78–9, 97, 104, 111, 208, 222
cluster analysis 132, 145–6
clusters; see regional clusters
collaboration; see networking
collective agents 202–6
collective enterprise, initiatives; see regional collective enterprise, initiatives
collective learning; see regional collective learning
competences 186–9, 195n, 232, 236–8
competition policy 237, 240, 252–3, 256–7
computer industry, especially software and services 4, 6–7, 36–8, 41–2, 44, 95, 111, 170, 209, 212, 234

constraints on high-technology SMEs 247–8
critical mass 51–2, 213, 215–6
cryogenics 107
cultural factors, influences 119–20, 131–2, 136, 162, 202–5
customers, especially local 16n, 123–4, 126–37, 142n, 149, 167, 173, 214–6, 242

defence expenditure and policy 60, 67, 244

economies of scale, regional 51–2
embedded laboratories 42–3, 51
endogenous growth processes 28, 40, 42, 67, 177
enterprising behaviour 8
entrepreneurs, entrepreneurship 7, 22, 42, 50, 58, 91, 98, 120, 159, 169–72, 202–3, 207–9, 213, 226n, 236, 250, 256
epistemic communities 183–5, 188, 195n
epistemically significant moments 189–95
ESRC Centre for Business Research 2, 15n, 46, 96, 103, 142, 219, 226n
EU funding 60, 105
EU technology policy 60
European Commission 2
evolutionary economics 232–6
external business advice; see business services, support, advice

factor analysis 145–51
failures, failure rates 39, 97
flexible specialisation 7, 24, 183
flexibility 28
foreign direct investment 176

German technology policy 60

Index

global links, networks 12–14, 27, 29, 163, 177, 210, 213, 217–8, 246
globalisation 13–14, 27, 29–30, 43, 47, 49–51, 159, 161, 163, 172, 174–5, 178, 183, 210, 213, 217–8, 246, 252, 257
Göteborg region 31, 34, 41–2, 64, 78–9, 96–7, 104–5, 167, 171–2, 208, 211, 215–6, 219, 221–2, 226n
government policies 7, 15n, 25, 57–72, 168, 230–57
government research laboratories; *see* research institutes, public sector
GREMI 15n, 119, 141, 142n, 188, 192, 200
Grenoble region 33–4, 67, 69, 71, 79–81, 98, 101, 104–10, 165–6, 173–4, 177, 204–6, 208, 212, 214–6, 222–3, 248–9

Hannover-Brunswick-Gottingen research triangle 127, 129
Helsinki region 38, 81–2, 97–8, 103, 112, 167, 204, 248, 250
Helsinki University of Technology 43, 81, 206
'high learning' regions 202–3, 209, 213–4, 218, 225, 246
high-technology growth 6–7, 15n, 41, 175, 250
high-technology industry 4, 235
high-technology services 38, 42, 169–70, 235
high-technology SMEs 4–8, 13, 15n, 30, 38, 41, 46–51, 92, 102, 158–63, 169–70, 219, 226n
high-trust relationships 50
highly-qualified entrepreneurs 50, 207
highly-qualified staff, labour 6–7, 34, 91, 209–13, 221, 247–8, 255
human capital 174

incubators 8, 64, 98, 108, 170, 221, 224, 250
industrial districts 16, 23–4, 118–9, 127–8, 141, 142n, 193
industrial regions 31–4, 163, 204, 242
information technology 41, 44, 89, 107, 170, 209, 215–6, 234
innovation 8–10, 14, 15n, 24, 27, 46–9, 51, 101–3, 121, 131, 140, 147, 158–9, 161–2, 174, 177–8, 191, 200, 203, 212, 217–20, 225, 226n, 237, 253, 256–7
innovation centres 64, 66, 75–88, 250
innovation systems 57, 237–40, 251, 254
innovative milieux 1, 10–12, 14, 29–30, 39, 66–7, 70–71, 91, 101, 118–22, 142n, 162, 166, 175, 188, 200, 206–10
institutional frameworks 15n, 57–72, 119, 246–9
institutional proximity 131–2, 137–8, 140–41, 142n
institutional thickness 72n
institutions, local and regional 11, 23, 57–72, 90–114, 220–22, 230, 247–9, 255
inter-SME linkages: *see* linkages, inter-SME
internationalisation: *see* globalisation
internet 43–4, 234, 238
inward investment 100–1, 161
islands of innovation 91
Italian high-technology milieux, regions 16n, 123, 127, 130, 136–40, 165, 214
Italian industrial districts 23, 28

knowledge centres 9, 14, 42–3, 64–6, 90–114, 134, 220–22, 241–2, 254
knowledge communities 199
knowledge workers (*see also* labour markets) 204, 210
knowledge-based economy 10–11, 183

labour force turnover 131–2, 137, 212
labour markets, especially local scientific, technical and professional 29, 35, 38, 43, 91, 102–4, 137, 145, 150–51, 200, 204, 209–13, 243, 248
labour mobility 188, 200–3, 209–13, 244
labour productivity 136–41, 154
large firm-small firm links: *see* links, large firm-small firm
large firms 5, 12–13, 15, 22–7, 29, 33–8, 42, 46, 50–51, 58, 66, 92, 156–78, 204, 207–10, 217, 226n, 243, 252–3, 256–7
learning preconditions 201–7, 220
learning processes 15n, 182–5, 187–8, 191, 199–226, 232–5, 243

learning regions 10–12, 187, 199
linkages, inter-SME 118–42, 203–4, 211, 214–8, 245, 248, 256
Linkoping 8, 15n, 245
Links, large firm-small firm 172–4, 177, 204, 243, 245, 252–3, 256–7
Localisation 27–8
Location advantages 150
Lock-in 31, 234, 239, 243, 247
Ludwig Maximilian university 33, 83, 99, 102, 166

market niches (*see also* niche markets) 7, 23, 44
metropolitan regions 30, 37–8, 164, 204
Milan region 37, 65, 82–3, 100, 136, 165, 208, 212, 216, 219
Minnesota 171
Munich region 33, 41, 46, 66–7, 71, 83–5, 95, 99, 101–2, 111, 166, 173, 216–9, 222, 226n, 244–5
multinational firms 12–13, 25, 36, 42, 66, 157, 159, 161, 163–8, 171, 173, 175–8, 212, 233, 243

Netherlands 7, 27, 95
networking, including collaboration (especially regional) 9–10, 12–13, 15n, 28–30, 42–3, 47–9, 120–23, 141–2, 161–2, 164, 172–4, 176–8, 192–3, 200–7, 214–8, 221–3, 232–4, 236–41, 243–4, 248, 252–3, 255
new firms (*see also* spin-offs) 7–8, 15n, 34, 158, 171, 203
new technology-based firms (NTBFs) 15n, 41–2, 44, 98–9, 158–61, 176, 209–10, 224, 226n, 243
newly-emerging technologies 3, 234, 239
niche markets (*see also* market niches) 47, 49, 165, 174, 215

organisational learning 185, 196n, 239
organisational proximity 131–2, 137–8, 140–41, 142n
Oulu 8
Oxford region 14, 31, 35, 42, 64, 66, 71, 87, 126–8, 131–2, 164–5, 172, 204, 208, 211–2, 215–6, 218–9, 222

Oxford university 35, 87, 96, 99–100, 103–5, 104–10, 215, 222, 226n

partnerships; *see* networking
patents 31, 138, 219, 237, 252–3, 257
path dependency 28, 31, 39
peripheral regions 92, 164, 245, 250
Piacenza 136, 165, 208, 212, 216, 219
Pisa region 35–6, 42, 100, 136, 165, 204, 208, 212, 216, 219
planning policies 249–50
preconditions for learning: *see* learning preconditions
professionals 7, 35, 204
public research institutes: *see* research institutes

quality of life 8–9, 213

R&D activity and expenditure 3, 24, 61–4, 138, 220–23, 235, 239, 241–6, 254–7
R&D consultancies; *see* research consultancies
R&D laboratories 51, 69, 100–1, 176, 212, 243
R&D-intensive firms 7, 46
Randstad region 38, 166–7, 171, 204
recession 15n, 25, 39
regional clusters 1, 8–9, 15n, 24, 27–30, 31–8, 39–45, 51, 70, 119, 121, 156, 159, 161, 176–8, 199–226, 249–50
regional collective enterprise, initiatives 43, 52, 95, 222–3, 246
regional collective learning 9–12, 15n, 30, 51, 120–21, 128, 131, 141–2, 161, 164–5, 182–95, 195–6n, 199–226, 230, 245–8, 251, 253–7
regional competences 187–9
regional convergence 41–3, 52n
regional development policies 249–51
regional disparities 230–31, 250–51, 255–6
regional governments 69
regional image 104–5
regional innovation systems 10–11, 57–72, 75–89, 252
regional sectoral structures, specialisation 41, 120, 169, 221, 242, 256
regulatory frameworks 237–8, 240, 257

research activity, expenditure; *see* R&D activity, expenditure
research collaboration with other firms 11–12, 14, 28, 120–21, 123, 126–7, 129, 132–4, 156, 161, 173–4, 177, 200–1, 215, 217–8, 252
research consultancies 42, 77, 99, 172, 203, 205, 209, 243–5
research institutes, public sector 75–88, 90–114, 127–8, 159, 166, 203, 205–9, 214–5, 222, 241–4, 249–51, 253–6
research laboratories; *see* R&D laboratories
residential amenity, attractiveness 8–9, 213, 222, 250
Rotterdam 102

satellite platform 42
science and technology regions 31–2, 62
science parks 8, 34–5, 37–8, 58, 75, 77–8, 81–2, 86–7, 96, 103, 105–6, 221, 224, 250
scientific instrument industry 176
sectoral structures, specialisation; *see* regional sectoral structures, specialisation
Siemens 66, 166, 173, 204, 216–7, 222
Silicon Valley 29–30, 49, 121, 164
small firms 5, 21–6, 39, 41, 44, 92, 119–21, 158–9, 171–2, 226n, 244, 253
SMEs 5–8, 21–5, 161, 169–71
social relations of innovation 236–7
socio-cultural factors, influences 15n, 119–20, 131–2, 136, 185, 202–5, 232, 240
Sophia-Antipolis 36, 39–43, 51, 64, 66–9, 85–6, 94–5, 99–100, 102, 104, 168, 176–7, 204, 208, 210, 212, 215–6, 222–3, 242–4, 250
spin-offs 8, 13, 25, 29, 34–6, 42, 64, 79, 92, 95, 97–100, 107, 131–2, 151, 156–78, 188, 192–3, 201–3, 205–9, 212, 215, 221–2, 242–3, 245, 249–50
subcontracting 23, 25, 46, 126–7, 204, 215, 217
subsidiaries 168, 173, 175, 177, 217
suppliers, especially local 16n, 123–4, 126–37, 142n, 148, 167, 173, 214–7
synergy 202, 253

tacit knowledge 27, 29, 120–21, 174–5, 182, 185, 189, 191–4, 212, 232–3, 243

takeovers 42, 174–6, 178
technological change 7, 24, 44, 50–51, 163, 183, 209, 234–5, 256
technology centres, parks 8, 38, 64–6, 69, 75–88, 101, 105–6
technology fairs 58, 66, 76–88
technology policies 58–72, 231–5, 251
technology transfer 13, 33, 35, 64–7, 72n, 90–114, 174, 177, 244, 249, 251, 257n
technology-based SMEs; *see* high-technology SMEs
technology-intensive SMEs; *see* high-technology SMEs
technopoles 164
telecommunications 7, 41–2, 111, 238
time dependency 204–7, 216, 250, 255
transactions costs 119, 141, 162–3
transnational firms; *see* multinational firms
trust 50, 119, 123, 148–9, 201, 214, 232, 238, 241, 247
TSER European network 2, 142n, 232

universities 9, 13, 15n, 24, 33–8, 42–3, 47–9, 58–72, 75–88, 90–114, 127–8, 159, 163, 167–8, 170–72, 193, 203–16, 218, 220–22, 241–3, 245, 248, 250–55
University of Nice/Sophia-Antipolis 43, 85, 104, 206, 208, 222, 242–3
university-based regions 34–7, 163, 171, 204
university spin-offs 97–100, 170–72
untraded interdependencies 12, 15n, 120, 199–200, 207, 209, 214
urban-rural shift 8
Utrecht region 88–9, 100, 102, 131, 134–6, 166–7, 171, 174, 177
Uusimaa county 38, 41, 167

venture capital 8, 169, 223, 225

West Midlands 251
Western Crescent 8
widely-diffusing technologies 3, 234, 239

ZIRST 34, 69, 80, 98, 105